Girum G. Alemu
Managing Risk and Securing Livelihood

ERDKUNDLICHES WISSEN

Schriftenreihe für Forschung und Praxis

Begründet von Emil Meynen

Herausgegeben von Martin Coy, Anton Escher und Thomas Krings

Band 159

Girum G. Alemu

Managing Risk and Securing Livelihood

The Karrayu Pastoralists, their Environment and the Ethiopian State

Franz Steiner Verlag

This work has been published with a kind support of BIGSAS

Umschlagabbildung: Karrayu camel herders near Nazreth/Adama town, Ethiopia.
Photo taken by Girum G. Alemu (2014)

Bibliografische Information der Deutschen Nationalbibliothek:
Die Deutsche Nationalbibliothek verzeichnet diese Publikation in der Deutschen
Nationalbibliografie; detaillierte bibliografische Daten sind im Internet über
<http://dnb.d-nb.de> abrufbar.

Dieses Werk einschließlich aller seiner Teile ist urheberrechtlich geschützt.
Jede Verwertung außerhalb der engen Grenzen des Urheberrechtsgesetzes
ist unzulässig und strafbar.
© Franz Steiner Verlag, Stuttgart 2016
Zugl.: Bayreuth, BIGSAS, Diss., 2014
Druck: Laupp & Göbel GmbH, Gomaringen
Gedruckt auf säurefreiem, alterungsbeständigem Papier.
Printed in Germany.
ISBN 978-3-515-11404-2 (Print)
ISBN 978-3-515-11405-9 (E-Book)

TABLE OF CONTENTS

LIST OF FIGURES 9

LIST OF TABLES 10

LIST OF PHOTOS 11

LIST OF ABBREVIATIONS 12

GLOSSARY 14

ACKNOWLEDGEMENTS 17

ABSTRACT 19

ZUSAMMENFASSUNG 21

1. INTRODUCTION 23
   1.1 The problematics of the hegemonic adaptation perspective 23
   1.2 Apolitical adaptation and pastoralists' development 25
   1.3 Approaching the 'local contexts' from a political ecology
       perspective 28
   1.4 Point of departure: risk, resources and relational modes 30
   1.5 Organization of the book 33

2. ADAPTATION, VULNERABILITY AND LOCAL AGENCY:
   THEORETICAL AND CONCEPTUAL REFLECTIONS 35
   2.1 Introduction 35
   2.2 Putting the concept of adaptation in perspective 35
   2.3 The conventional approach to adaptation 36
   2.4 The inadequacy of the conventional approach 38
   2.5 The concept of vulnerability in adaptation research 39
       2.5.1 Socially embedded vulnerability 41
       2.5.2 Not always passive victims: focus on the locals' agency 43
   2.6 Political ecology and the vulnerability paradigm 45
   2.7 Locating actors in political ecology 47
   2.8 Summary 50

3. RESEARCHING WITH THE LOCALS: METHODOLOGICAL
   REFLECTIONS 51
   3.1 Introduction 51
   3.2 Upper Awash valley, Fentalle *Woreda*: descriptions 52

        3.2.1 Research area and the people 52
        3.2.2 Physical features 54
        3.2.3 Micro-climate and seasons 55
    3.3 Methodological stances: qualitative approach in geography 56
        3.3.1 Qualitative approach 56
        3.3.2 Participatory approach 57
    3.4 The research process: preliminary visits and initial acquaintance 58
        3.4.1 Selecting the units of analysis and study villages 58
        3.4.2 Specific research methods employed 60
    3.5 Mixing various interview techniques 63
    3.6 Field notes and participant observation 66
    3.7 Document analysis and secondary information 67
    3.8 Analysis and write up 67
    3.9 Ethical considerations 68

4. LIVELIHOOD INSECURITY IN CONTEXT: HISTORICAL
   TRAJECTORIES 70

    4.1 Introduction 70
    4.2 State-pastoralists relations: 'development' and the perils of
        planning 71
        4.2.1 Commercial farms and loss of access to resources 71
        4.2.2 Conservation without people 74
    4.3 Environmental sources of livelihood risk 79
        4.3.1 Water scarcity, frequent drought and food insecurity 79
        4.3.2 Pastureland degradation and encroachment of invasive
              plants 85
    4.4 The post-1991 state in pastoral spaces: a liability or an asset? 87
        4.4.1 Political reconfiguration and the recognition of a pastoral
              way of life 87
        4.4.2 Transforming the nomads: rationality of state development
              projects 89
    4.5 In-migration and increased population pressure 94
    4.6 Summary: multiple sources of livelihood insecurity 97

5. ENVIRONMENTAL TRANSFORMATION AND LIVESTOCK-BASED
   LIVELIHOOD PRACTICES 100

    5.1 Introduction 100
    5.2 Environmental transformation and social disorganization 100
        5.2.1 Distortion of settlement and mobility patterns 101
        5.2.2 Changes in the social fabric 104
        5.2.3 Disruption in customary resource management institutions ... 105
    5.3 Pastoralists' agency and response to change 109
    5.4 Contexts influencing pastoral mobility decisions 114
    5.5 Reorganization: camel-based livelihood practices 117

5.6 Surviving on camels: risk management and livelihood practices 118
　　　　5.6.1 Knowledge and herd management 119
　　　　5.6.2 Access to browse: the routes of camel mobility 123
　　5.7 Summary 127

6. LIVING THE TRANSFORMATION: THE MOVE TOWARDS AGRO-PASTORALISM 128

　　6.1 Introduction 128
　　6.2 Cultivation as risk management and livelihood practice 128
　　6.3 Significant aspects influencing decisions to cultivate 130
　　　　6.3.1 Disruption of social relations and increase in farming 130
　　　　6.3.2 Population pressure, settlement and farming practices 134
　　　　6.3.2 Policy factors: irrigation and land certification 136
　　6.4 Bringing new principles in: fencing communal pasture 141
　　6.5 Emergence of new arrangements 142
　　　　6.5.1 Benefit-sharing arrangements 145
　　　　6.5.2 Resource-sharing arrangements 146
　　6.6 Some constraints to cultivation as a contemporary strategy 150
　　6.7 Non-pastoral and non-agricultural activities 152
　　6.8 Summary 156

7. CONTEXTUALIZED ADAPTATION: HEGEMONIC PERSPECTIVES AND LOCAL RESPONSES 158

　　7.1 Introduction 158
　　7.2 The politics of adaptation: top-down approaches to development 159
　　　　7.2.1 Consideration of governance structures in adaptation 159
　　7.3 Situated and local-level adaptation practices 161
　　　　7.3.1 Recognition of local agency 163
　　7.4 Summary 167

8. CONCLUSIONS 168

　　8.1 Starting-point vulnerability: the political ecology of local
　　　　adaptation 168
　　8.2 Locating agency in political ecology 170
　　8.3 Localizing the international perspective on adaptation to climate
　　　　change 172
　　8.4 Back to the research questions 173
　　8.5 Implications for policy and directions for future research 175
　　　　8.5.1 Addressing social sources of vulnerability 176
　　　　8.5.2 Increasing the negotiation power of the locals 176
　　　　8.5.3 Directions for future research 177

9. LIST OF REFERENCES 179

LIST OF FIGURES

Figure 1: Interpretation of vulnerability in climate change adaptation research ...41

Figure 2: Local actors in a broader political-ecology context 48

Figure 3: Map of the study area showing Fentalle *Woreda* 53

Figure 4: The short season rainfall pattern of Fentalle *Woreda* (1985–2009).......55

Figure 5: Cultivated area by crop variety in Upper Awash Valley (1970s) 73

Figure 6: Dry-season pasture land lost to various interventions (in hectares).......76

Figure 7: The progression of entitlement failure of pastoralists............................77

Figure 8: Annual growth of profit (in Birr) in the Metehara Sugar Factory..........78

Figure 9: The annual rainfall pattern of Upper Awash valley (1966–2004) 80

Figure 10: Number of PSNP beneficiaries in Fentalle *Woreda* (2007–2009).......82

Figure 11: Major trends, events and responses in Fentalle *Woreda* 84

Figure 12: Trends in population number of the Karrayu since 1968....................95

Figure 13: Migration routes of the Ittu to Karrayu land 96

Figure 14: Summary of the context of livelihood insecurity.................................99

Figure 15: New forms of mobility by camel herding Karrayu pastoralists 124

LIST OF TABLES

Table 1: Pair-wise ranking of major sources of livelihood insecurity 83

Table 2: Area recommended for forage development user households 93

Table 3: Major seasons and mobility destinations in Fentalle *Woreda* 102

Table 4: Gendered and age-based roles within Karrayu pastoral households 113

Table 5: An example of camel categorization in the Karrayu community 120

Table 6: Camel herders' roles and responsibilities by age 122

Table 7: Summary of mobility types as practised by the camel-herders 126

Table 8: The different practices and mechanisms for securing resources 148

Table 9: Farming households' responses to constraints on crop production 151

Table 10: Number of Karrayu individuals employed in the MSF 154

Table 11: The pastoralists-government understanding of adaptation 166

LIST OF PHOTOS

Photo 1: Key informant interviews with Karrayu elders 64

Photo 2: Partial view of the Metehara Sugar Factory and plantation 74

Photo 3: Degraded pasture in the Karrayu rangeland 86

Photo 4: The practice of farming in Karrayu community 131

Photo 5: The new irrigation scheme has increased water availability 138

Photo 6: Fencing off communal grazing areas for farming purposes 139

Photo 7: Land preparation by Ittu farmers 149

Photo 8: Selling of fuel-wood and charcoal in Metehara town 155

LIST OF ABBREVIATIONS

ANP	Awash National Park
CBDRR	Community Based Disaster Risk Reduction
CRGE	Climate Resilient Green Economy
CSA	Central Statistical Authority
DA	Development Agent
DPPC	Disaster Prevention and Preparedness Commission
EMA	Ethiopian Metrological Agency
EPRDF	Ethiopian People's Revolutionary Democratic Front
FDRE	Federal Democratic Republic of Ethiopia
GDP	Gross Domestic Product
HVA	Hendels-Vereeiging Amsterdam
IPCC	Inter-governmental Panel on Climate Change
MSF	Metehara Sugar Factory
NAPA	National Action Plan for Adaptation
NOPA	Network of Pastoralists Association
PASDEP	Plan for Accelerated and Sustained Development to End Poverty
PRSP	Poverty Reduction Strategic Paper

PSNP	Productive Safety Net Program
SDPRP	Sustainable Development and Poverty Reduction Paper
UN	United Nations
UNDP	United Nations Development Program
UNFCCC	United Nations Framework Convention on Climate Change

GLOSSARY

Abuuruu: scouts who supervise pasture and water resources available in a grazing land before migration takes place.

Khat: A kind of plant that is widely grown in Harar and South Western parts of Ethiopia and used as a stimulant

Damina: A traditional *gosa* (clan) leader elected by the council of elders of that particular gosa

Derg: Amharic term which literally means committee

Gada: The central institution in the Socio-political way of life of the Oromo. It is a system of classes (sets) that succeed each other every eight years in assuming power and offering political, economic, military and ritual leadership to the Karrayu as well as other Oromo people especially in the period before the external interventions.

Ganda: The smallest unit of local organization that makes reference to a territory in the Karrayu traditional organization

Kebele: The minimal structure of government administration unit, countable to the district. It is introduced during the *Derg* rule in Ethiopia.

Kalloo: The practice of reserving pasture for dry-season among the Karrayu

Malkaa: Watering points along the Awash River. They are also holy grounds and places of *gada* power transfer.

Salfaa: Armed guards that the *gada* leaders recruit and send to protect the Karrayu pastoralists especially in their migration to Haroole and Bulga/Kessem River from hostile neighbours.

Waaqeffataa: One who believes in the Oromo traditional religion called Waaqef-fannaa which adheres to the belief in one God called Waaqaa.

ACKNOWLEDGEMENTS

Many people have supported me in the process of writing this thesis. Supervisors, research participants, family, and friends have provided important information, advice and support, without which its completion would have been impossible. Here, I wish to thank a few people in particular.

Firstly, I would like to thank my supervisor, Professor Dr. Detlef Müller-Mahn, for his intellectual guidance. Professor Müller-Mahn's endless intellectual capacity intertwined with his personal commitment and caring abilities made this journey so interesting and lively. I also admire his sixth sense. He knows when the student needs a push, needs a break, and above all he is able to coach students to discover their own potential. Thank you very much. Additionally, I would like to recognize the contributions of my mentoring members, Professor Dr. Martin Doevenspeck and Professor Dr. Kurt Beck. The critical comments and suggestions of my external examiner, Dr. Ayalew Gebre, were also instrumental in improving the final version of this work.

I would also like to take this opportunity to thank all the staff members of the department of social and political geography at the University of Bayreuth for the wonderful working atmosphere and lively discussions on several occasions. In particular, I benefited a lot from discussions with Dr. Jonathan Everts. Special thanks to Michael Wegner for patiently preparing the maps and graphics. Michael, thank you very much for your brilliance in changing the ideas I told you into convenient graphs and maps. I would also like to thank Dr. Ruth Schubert for her professional editing of the manuscript.

I received a great deal of financial and institutional support for this research. I owe my deepest gratitude to the German Academic Exchange Program (DAAD) for the scholarship that I was granted to pursue my studies. I also thank the Bayreuth Graduate School of African Studies (BIGSAS) for their generous support of my field work. They also covered my travel costs for attending conferences. Without your financial commitment this work would never have been possible. Thank you very much.

The encouragement and support from my family has always been with me throughout my studies. My beloved wife, Johanna Alemu, is a great partner who always earnestly encourages me and enthusiastically follows up the progress of my work. She is, indeed, typical of the strong lady behind her partner's success. Jo, thank you so much for all your support!

Parents and relatives at home and here in Germany also had their role to play in the success of my study. I thank my parents – Getachew Alemu, my father, and Alemitu Dhugo, my mother – above all for giving me the chance to learn and who eagerly anticipated the successful completion of my studies. I am indebted to my parents-in-law, Manfred and Claudia Knoll, and all their family members, for their warm embrace and making me feel at home here in Germany.

ABSTRACT

This work explores how pastoralists-environment-state interactions influence risk management and livelihood practices in dryland pastoral areas of upper Awash valley, Ethiopia. It is informed by a case study of the Karrayu pastoral community in Fentalle *Woreda*. Climate change poses particular challenges for Karrayu pastoralists, who, in general, are largely natural-resource dependent, dwelling in arid and semi-arid environs and experiencing high climate variability and extremes. This study responds to the lack of critical attention paid to local responses to climate-related stresses which are closely interwoven into the social fabric of the pastoralists. The research addresses the gaps in mainstream international adaptation knowledge, thereby contributing to an inclusion of local mechanisms for dealing with risk and livelihood insecurity and better integration of pastoralists' knowledge in national programmes.

By locating the research in a social geography tradition, I explore dominant understandings of vulnerability and adaptation and how these have implications for local level traditional responses in the Karrayu pastoral context. The study is informed by qualitative research based on the critical political ecology approach, which puts emphasis on power differentials and the agency of the locals in shaping their risk management and livelihood practices. A combination of semi-structured interviews, unstructured interviews, participatory techniques and observation were used to investigate the ways in which the locals construct their vulnerability and adaptation needs in dryland areas of Upper Awash Valley.

Throughout this research it has been possible to show that there is a divergence between the local construction of vulnerability and the dominant understanding of it in the international climate change perspective. This has implications for the adaptation practices envisaged by the locals and on the national and international levels as recorded in various documents. The Karrayu pastoralists construct vulnerability to climate-related stress as being caused by predominantly social factors and processes. Accordingly, the local arrangements for responding to the negative implications of a range of climate stresses and uncertainties are integrated into everyday risk management and livelihood practices. However, the traditional mechanisms of risk management and livelihood practices are threatened by social changes stemming from broader non-local processes. The pastoralists are using their agency to respond to these new challenges and powerful structural forces by creating new mechanisms for reorganizing their scarce resources. They are actively responding to the changing environmental context, rather than being passive agents. It is clear that any official adaptation measures meant to improve pastoral agency require local level development initiatives, targeting social processes which are at the core of increased pastoral vulnerability. However, in dominant international perspectives, vulnerability to climate change is regarded as being caused by specific climate-related stimuli, their biophysical impacts and the ability to directly respond to these. Consequently, interventions in dryland pastoral areas are characterized by technical measures that reactively respond to particular climate impacts rather than proactively reducing vulnerability. In this study I argue that the dominant adaptation perspective limits our understanding of the locally ingrained mechanisms of risk management and livelihood practices in the Karrayu pastoral community, where the causes of vulnerability are embedded in broader social contexts.

ZUSAMMENFASSUNG

Diese Arbeit untersucht, inwieweit Interaktionen zwischen Hirtengemeinden, Umwelt und Staat das Risikomanagement und Maßnahmen zum Erhalt der Lebensgrundlagen in Trockengebieten der Upper Awash Valley, Äthiopien beeinflussen. Sie basiert auf einer Fallstudie der Karrayu Hirtengemeinde in Fentalle *Woreda*. Der Klimawandel stellt besondere Herausforderungen für die *Karrayu* Hirten dar, die in der Regel in ariden und semi-ariden Gebieten mit einer hohen Klimavariabilität und starken Klimaextremen leben, und deren Lebensgrundlage weitgehend von natürlichen Ressourcen abhängt. Anhand dieser Studie werden lokale Antworten auf klimabedingte Belastungen untersucht, die eng mit dem Sozialgefüge der Hirten verwoben sind, und denen oft viel zu wenig Beachtung geschenkt wird. Die Recherche adressiert somit Lücken im Mainstream-Diskurs zu internationaler Anpassung und trägt zur Einbindung lokaler Mechanismen bei, die als Ziel haben, Risiken und Unsicherheit der Lebensgrundlagen zu reduzieren und eine bessere Integration des Wissens der Hirtengemeinden in nationalen Programmen zu erreichen.

Durch die Einordnung der Recherche in eine sozialgeografische Tradition, erforscht diese Studie das vorherrschende Verständnis von Vulnerabilität und Anpassung und welche Auswirkungen diese für traditionelle Ansätze auf lokaler Ebene im Kontext der Karrayu Hirtengemeinde haben. Die Studie stützt sich auf qualitative Recherche basierend auf einem politisch-ökologischen Ansatz, der Machtverteilungen/Machtgefälle und die Fähigkeit lokaler Akteure, ihr Risikomanagement und Maßnahmen zum Erhalt ihrer Lebensgrundlagen selbst zu bestimmen, als Schwerpunkt setzt. Eine Kombination aus semi-strukturierten Interviews, unstrukturierten Interviews, partizipativen Methoden und Beobachtung wurden angewandt um Möglichkeiten zu untersuchen, wie lokale Akteure ihre Vulnerabilität und Anpassungserfordernisse in den Trockengebieten der Upper Awash Valley definieren.

Durch diese Forschung konnte gezeigt werden, dass eine Abweichung zwischen der lokalen Konstruktion und dem vorherrschenden Verständnis von Vulnerabilität aus der internationalen Klimaschutzperspektive besteht. Dies hat Auswirkungen auf die Anpassungspraktiken, die von lokalen Akteuren und auf nationaler und inter-nationaler Ebene angestrebt werden, und in verschiedenen Dokumenten aufgezeichnet sind. Die Karrayu Hirtengemeinde definiert Vulnerbilität für klimabedingte Belastungen als überwiegend durch soziale Faktoren und Prozesse verursacht. Dementsprechend werden lokale Maßnahmen, auf die negativen Auswirkungen einer Reihe von Klimabelastungen und unsicherheiten zu reagieren, in das tägliche Risikomanagement und in Praktiken zum Erhalt der Lebensgrundlagen integriert. Allerdings werden die traditionellen Mechanismen des Risikomanagements und zum Erhalt der Lebensgrundlagen durch gesellschaftliche Veränderungen bedingt durch breitere, nicht-lokale Prozesse bedroht. Die Hirten nutzen ihre Handlungsfähigkeit, um auf diese neuen Herausforderungen und mächtigen strukturellen Kräfte zu reagieren, indem sie neue Mechanismen entwickeln, um ihre knappen Ressourcen zu reorganisieren.

Sie reagieren somit aktiv auf Veränderungen in ihrer Umwelt, anstatt passive Akteure zu sein. Zunächst ist klar, dass jegliche offiziellen Anpassungsmaßnahmen mit dem Ziel, die Handlungsfähigkeit der Hirtengemeinden zu stärken, lokale Entwicklungsinitiativen erfordern, die gesellschaftliche Prozesse anvisieren, welche im Zentrum der Vulnerabilität der Hirten stehen. Hingegen stehen klimabedingte Stimuli, ihre biophysikalischen Auswirkungen und die Fähigkeit, direkt auf solche Auswirkungen zu reagieren, im Zentrum vorherrschender internationaler Ansichten zum Thema Klimavulnerabilität. Folglich zeichnen sich Interventionen in ländlichen Trockengebieten durch technische Maßnahmen aus, die auf bestimmte Klimaauswirkungen reagieren, anstatt proaktiv Vulnerabilität zu reduzieren. Diese Studie argumentiert, dass die vorherrschende Ansicht im Anpassungsdiskurs das Verständnis lokal verwurzelter Mechanismen für Risikomanagement und für den Erhalt der Lebensgrundlagen einschränkt, wie zum Beispiel in der Karrayu Hirtengemeinde, wo die Ursachen der Vulnerabilität im breiteren gesellschaftlichen Kontext eingebettet sind.

1. INTRODUCTION

1.1 THE PROBLEMATICS OF THE HEGEMONIC ADAPTATION PERSPECTIVE

Adaptation to climate change has become a hegemonic concept at global level. This hegemonic discourse not only influences research agendas but also directs the way development is being practised. This global mainstream perspective on adaptation is founded on special sets of concepts and practices that form the general context in which the term adaptation directs particular activities at local levels. Within the international community, adaptation is articulated on the basis of specific understandings of the term that shape the entire process of formulating and disseminating knowledge surrounding the concept. In this regard, the international adaptation policy is the crux of the mainstream international adaptation perspective. Interpretations of it influence central avenues of comprehension in respect of vulnerability and adaptation and the way that it is put into practice in adaptation implementation.

There is a widely accepted consensus and growing concern that global climate change is triggering significant social and ecological transformations that will heavily jeopardize the livelihood of populations in developing countries. Many of the developing countries are considered as the most vulnerable to climate change. Accordingly, with regard to addressing the issue of climate related risks, developing and developed countries have been assigned collective but distinct responsibilities under the United Nations Framework Convention on Climate Change (UN, 1992: Article 1). In more specific terms, developed countries are expected to support the adaptation activities implemented by developing nations as part of ongoing development assistance and cooperation (UN, 1992: Article 4(4)). Even though there is an increase of funding to financially support adaptation measures undertaken in developing countries, there remains a gap in understanding how these adaptation measures are directed to support the needs of the locals in developing countries.

Although climate change, manifested through environmental hazards, is a 'real' environmental issue 'out there', the way it is portrayed and responded to is socially constructed, being dependent on 'social frames'. In other words, while the predominant drivers of climate change operate within the broad scales of the global politcal economy, the aggregate impacts of climate change are most felt at the household and community levels. Within development policy and academic circles it is widely accepted that climate change is an environmental problem arising from atmospheric emissions. The depiction of the problem of climate change as arising from the accumulation of greenhouse gases in the atmosphere, and the consequent biophysical hazards, has completely diverted attention from problems embedded in social, political and economic structures. The problem of climate change could have been formulated in other ways by emphasizing the structural forces of the capitalist economy as the major driver of greenhouse gas emissions and thereby disruptions of livelihood and destitution (Demeritt, 2001). This line of argument brings to light the partiality of the scientific construction of climate change. The scientific depictions of climate change impacts, responses and causes are shaped by different discourses. The scientific understanding of climate change impacts neglects other crucial concerns, such as inequality and development, when seeking suitable responses at various levels (O'Brien and St. Clair, 2007; Liverman, 2009).

The dominant perspective on adaptation that gives primacy to climate stimuli underestimates processes of development that impact livelihoods in developing countries. Such an underestimation also influences our focus and understanding of risk management and livelihood responses that are mainly developed through the communities' interactions with their environment. A sole focus on climate stimuli cannot capture politically, socially, institutionally and legally complex responses.

Because the natural environment is seen as the root source of the climate change problem, the responses to this problem are also framed as a scientific endeavour that shields humanity from wild nature (Cass and Pettenger, 2007). This understanding of climate change, which gives primacy to biophysical elements while neglecting social and political factors, underscores nature as the main source of danger. This leads to the understanding of vulnerability as an outcome of biophysical factors (Gaillard, 2010). This is clearly a result of the dichotomy between science and politics. This dichotomy is also reflected in the representation of the Intergovernmental panel on climate change (IPCC) as an 'upstream' independent organ with the mandate of compiling scientific solutions, which feeds information to politicians and decision makers located in the 'downstream policy process' (Demeritt, 2001). The constructed distinction between science and politics and between fact and value is entrenched in the climate change sphere. Science and politics, however, are mutually linked (Forsyth, 2003; Cass and Pettenger, 2007; Demeritt, 2001; 2006). Furthermore, the dominant framing of climate change adaptation puts biophysical environmental factors at the centre of community problems. One cannot question the influence of the environment on the risk management and livelihood practices of local communities. However, the assumption that climate change is a natural problem has led to the de-politicization and de-contextualization of both the climate change problem and the solution to it. In relation to the emergence of a decontextualized dominant climate change discourse, Forsyth (2003: 14) argues that the framing of an "environmental explanation...may be based on...norms of one society" and this environmental knowledge has been reproduced in other contexts where values and norms differ.

Even though the vulnerability paradigm is gaining ground in the fields of adaptation research, the hegemonic international perspective on adaptation still maintains an event-centred framework of adaptation and vulnerability that deals with the actual or expected outcomes of climate change, including variability and extremes (Schipper, 2007). In general, the hegemonic adaptation perspective frames adaptation activities as apolitical processes that are tied to actual or expected climate stimuli and hence these adaptation activities are isolated classifiable strategies independent of social processes of risk management and livelihood practices. Such a perspective has widely influenced development practices implemented by national governments by diverting interventions away from addressing the root causes of societal problems and solely focusing on dealing with climate stimuli. When this focus on distinct adaptation activities translates into development interventions, it completely misses the point that many of the environmental problems that rural communities experience emanate from the political economy and historical trajectories of development interventions.

1.2 APOLITICAL ADAPTATION AND PASTORALISTS' DEVELOPMENT

The way particular regions in the world are imagined is a result of cultural orientations towards these regions, which in turn also determine the construction of vulnerability in these regions (Bankoff, 2001; 2004). In the context of disasters, Bankoff contends that 'vulnerability' itself is a discourse, related to discourses of 'development' and 'tropicality' that sustain a Western, hegemonic perception of regions (like sub-Saharan Africa; East Africa; arid and semi-arid regions) as 'more dangerous' than the temperate West. Commitment to the mainstream knowledge system regarding climate change and adaptation generates generalizations about pastoral communities which, to some extent, predetermine the nature of vulnerability assessments and thus determine adaptation trajectories. The understanding of vulnerability within the climate change adaptation realm has a particular conceptual framework that is sustained by a dominant discourse of adaptation as something that is distinct from development or disaster risk reduction. A science and impacts focus generates generalizations to the effect that environmental factors are the primary contributor to vulnerability in arid and semi-arid areas. From a biophysical science and impacts perspective, it is easy to see why these areas may be perceived as vulnerable; they are highly susceptible to climate variability and extremes which are exacerbated by climate change.

Ethiopia, with its more than sixty per cent arid and semi-arid land mass, is categorized as being particularly vulnerable to climate change by the United Nations framework of conventions on climate change (UNFCCC) (see particularly Article 4(8c), UN, 1992) and the IPCC (Boko *et al.*, 2007). The issue of climate change compounds these notions of extraordinary vulnerability. In popular science and the media, pastoral groups residing in arid and semi-arid environments are depicted in an apocalyptic manner as 'climate canaries' and 'the people most likely to be wiped out by devastating global warming' (Observer, 12.11.2006 cited in Morton, 2010). Pastoralists who reside in arid and semi-arid environments like the Karrayu in the Upper Awash Valley are widely pictured as extremely vulnerable. These prevalent pictures emanate from broader economic and geographic imaginings of the region as ridden with conflict, drought, famine, hunger, suffering and death. These constructs are further legitimized by Western media reports that give only a snapshot view of pastoral areas, as in the following case: "They have long lived on the margins, a way of life that was manageable as long as the rains were regular. But with relentless drought the margins are coming close to being impossible" (Fergal Keane, BBC, 17.11.06 cited in Morton, 2010).

The dominant framing of adaptation as a means of managing climate risks that solely focuses on climate change impact reduction seriously hampers vulnerability reduction. There is recognition of the concept of vulnerability in international climate change adaptation research. However, event-centred approaches to vulnerability and adaptation based on present and expected impacts of climate change still dominate the international climate change adaptation discourses (Schipper, 2007). The dominant framing of adaptation and vulnerability in climate change research sees adaptation as an apolitical intervention based on scientific solutions to environmental problems. The mainstream adaptation discourse, as a result of its discrete policy and funding agenda and disciplinary roots, remains primarily science and impacts focused. This has particular consequences for vulnerability reduction in the dryland areas of East Africa in general, and Ethiopia in particular, as it may serve to facilitate the further implementation of ill-planned policies for pastoral areas. Bankoff (2001) cautions that "commitment to a particular knowledge system ... predetermines the kinds of generalisations made about the subject under investigation ..." (Bankoff, 2001:29). The overemphasis on actual and expected climate impacts within the dominant discourse has also led to the assumption of adaptation as identifiable strategies irrespective of the broader social context and additional to development and disaster risk reduction. In situations where the influence of climate-related stresses surpasses the ability of local practices and knowledge to respond and deal with them, adaptation measures become necessary to tackle particular climate change impacts, such as drought. Conversely, in line with Schipper (2007), I contend that in the context of Ethiopian pastoralists in the semi-arid parts of the Upper Awash Valley, reducing vulnerability, which is framed by the locals as a politico-ecological problem emanating from the political economy of development, has little to do with minimizing the potential for negative climate change impacts. Hence, the reduction of vulnerability by transforming the social conditions that govern resource acess, allocation and management is a much more difficult and complex task than minimizing expected or actual climate change impacts in this context. However, due to the overemphasis given to climate change impacts, official funding and implementation of adaptation activities usually ignores social contexts, and is reduced to actions to tackle specific impacts of climate stresses. Thus, the 'adaptation to impacts' approach in climate change research is a reactive response to known climate impacts, rather than a proactive response to vulnerability that tries to address the broader political and ecological context.

In this regard, development interventions in pastoral areas are still fixed on technological and scientific solutions, at the expense of a social science oriented approach to generating knowledge about climate-related problems. A particularly prominent outcome of this in pastoral areas is a perceived need for scientific certainty based on equilibrium assumptions, in order to proceed with adaptation for a very unpredictable and non-equilibrium environment. I do not deny the need for climate science research – this knowledge is always required to better understand the nature of future climate change challenges. However, lack of scientific knowledge and certainty is not what is hindering effective adaptation for the pastoralists in the dryland parts of Awash Valley. This prevailing perception detracts from the type of research – mainly social science research – that is urgently required if locally oriented adaptation trajectories are to produce effective outcomes for the pastoralists themselves. Improved scientific knowledge is not necessarily a priority for adaptation in the arid and semi-arid regions of the country; what is already known by 'experts' and local people are largely sufficient to proceed with effective adaptive actions. In order to better deal with environmental uncertainty, adaptation should be anchored in a sound local knowledge of disaster reduction, natural resource management and development. Though biophysical factors have an influence on local vulnerability, the overemphasis on geographical location leads to simplistic constructions concerning dryland areas and pastoralists which are overly pessimistic. This partial construction underemphasizes consideration of the people living in this dryland environment and their resilience, capacity, knowledge and agency; pastoralists possess a considerable capacity to cope with change and uncertainty. However, the neglecting of these capacities has led to displacement of the traditional mechanisms for handling vulnerability.

The tensions between national and local conceptualizations of vulnerability are historical and strong in pastoral areas of Ethiopia. For instance, the national discourses as they are reflected in various government documents consider the pastoralists' way of handling vulnerability as 'archaic', so that they need to be replaced with discrete interventions such as 'modern' settled agriculture. In many arid and semi-arid parts of the country, environmental challenges linked to climate change are indeed highlighting high levels of vulnerability. Climate change poses serious biophysical environmental challenges. However, pastoralists have been facing environmental uncertainty for generations and have a strong cultural tradition for dealing with it. I argue that the dominant scientific discourse of climate change vulnerability in dryland pastoral areas places an overemphasis on biophysical stresses at the expense of recognizing the importance of the socio-cultural, socio-economic and political factors creating vulnerability. The broad assumptions made about the nature of the challenges faced and solutions to them in pastoral areas, discourages consideration of socio-cultural capacity, resilience and agency. This shows the inability of biophysical causation to explain the impact of climate change. However, in this book, in line with recent studies such as that by Catley, Lind and Scoones (2013), I argue that pastoral areas are interlinked across space by the communities that live there, thus requiring a different approach to the meaning of development (and therefore adaptation). As insights from my study of the Karrayu pastoralists show, the trans-spatial nature of resources in dryland areas is a potentially valuable intrinsic component of resilience in the context of climate change and environmental uncertainty.

1.3 APPROACHING THE 'LOCAL CONTEXTS' FROM A POLITICAL ECOLOGY PERSPECTIVE

An understanding of local level and context-specific risk management and livelihood security has been approached from various disciplinary perspectives in the fields of sociology, geography, anthropology and development studies, using various perspectives such as sustainable livelihoods, natural resource management and disaster risk reduction. Each of these fields brings its own theoretical and conceptual framework and accordingly, there are many conceptual and methodological entry points to understand the way local people experience risks and how they manage these risks to their livelihoods. The perspective of disaster risk reduction, for instance, would approach adaptation as an amendment and build-up of local level activities so as to minimize the influence of vulnerability connected with hazards. On the other hand, the sustainable livelihoods perspective emphasizes strengthening local level resilience under conditions of climate-related shocks. Thus, clarifying the disciplinary orientation of this study is crucial before going further. In my research I have used the political ecology approach in order to understand risk management and livelihood security of communities at the intersection of society-environment-state interactions at the local level. This perspective helps us to locate the everyday livelihood and risk management practices pursued by rural people within the broader perspective of state interventions in the peripheral areas and the associated environmental transformations. Interactions between pastoralists and the state since the mid-20th century are the major historical factors that have laid the foundation for how today's pastoralists relate to and interact with their major resources – land and water. On the other hand, pastoralists-environment interaction in a non-equilibrium context is, and has always been, an ever-present phenomenon that plays a significant role in influencing the way livelihoods and risk management practices are pursued. In this regard, the variable and unpredictable arid and semi-arid environments in which pastoralists dwell requires a risk management and livelihood strategy that takes these variations into consideration. However, the pastoralists-environment interactions have been put out of balance through the historical trajectories of state development interventions that have resulted in different relational modes (property arrangements) to natural resources. State interventions in pastoral areas have not only disrupted pastoralists' relational modes to the natural resource base but have also brought in new resources that require societal reorganization and institutional rearrangements in order to adjust to the transformed environment. These disruptions in pastoralists-environment interactions has also resulted in increased susceptibility to the impact of drought, an increase in environmental degradation and food insecurity, and hence heavy dependence on outside support to sustain their livelihood and manage risks. Furthermore, the increase in frequency of extreme climatic variability will also likely compound many existing social problems.

Essentially, political ecology is a clear alternative to conventional 'apolitical' ecology (Robbins 2012). Political ecology, embedded in political economy, and to some extent in critical theory, developed in reaction to what were perceived as narrow and deterministic views on socio-environmental relations and change, mostly with respect to issues of power (Paulson and Gezon, 2005; Blaikie, 2008). Despite encompassing a variety of theoretical and methodological orientations, scholars in political ecology share a set of assumptions and viewpoints, which have also oriented this work. A central assumption is that "politics is inevitably ecological and that ecology is inherently political" (Robbins, 2012:3). This understanding of societal and ecological processes as being fundamentally interwoven calls for an integrated analysis of social and material aspects of environmental change. For that, political ecologists emphasise the value of place-based research and methodological pluralism (Paulson and Gezon, 2005). Additionally, by focusing on how political-economy systems and relations influence, and are influenced by, the environment and resources, researchers in political ecology elucidate the importance of multi-scale analysis (Bailey and Bryant, 1997; Robbins, 2012). Accordingly, drivers of environmental problems are often approached and contextualized in the larger political and economic context rather than "blamed on proximate and local forces" (Robbins, 2012:13) such as population growth or inappropriate resource management practices. Likewise, by focusing on access to and control over resources and social relations of production, political ecology illuminates the many and crucial tensions and conflicts in strategic interests, experiences, knowledges and practices among and between individuals and groups socially differentiated by overlapping relations of power rooted in gender, wealth and ethnicity (Rocheleau, *et al.*, 1996; Paulson and Gezon, 2005).

The political ecology approach that pays attention to state-pastoral-environment interactions refines our understanding of risk management and livelihood practices. Accordingly, this study focuses on the intersections of the contextual sources of environmental change, conflict over resources and the political consequences of environmental change. In particular, anchoring risk management and livelihood security at the intersection of environmental resources and the associated relational modes (access, institutions and strategies) contributes to a better understanding of contextualized responses to changes in environment-pastoralists relations. However, as has been argued in section 1.1 above, in the field of climate research the problem of detaching nature (the natural environment) from societal process has seriously misguided the formulation of international adaptation policies and implementations at the local level. There is a consensus that both at the international policy level and at the local implementation level, there is a need to incorporate climate change adaptation as well as disaster risk reduction into broader development practices so as to reduce vulnerability. There have been continuous debates on how to join up these apparently separate issues (Schipper and Pelling, 2006; Schipper, 2007; Schipper, 2009; Gaillard, 2010). By approaching risk management and livelihood security through a political ecology lens, this dissertation contributes theoretically to the debates of nature-society research, and practically to the fields of development policy. In light of the foregoing, I formulate and frame the research problem in its

1.4 POINT OF DEPARTURE: RISK, RESOURCES AND RELATIONAL MODES

Pastoralism as a way of life and a source of livelihood in the arid and semi-arid lowlands of Awash Valley is rapidly changing as land resources are becoming increasingly scarce and as herders increasingly have to confront new Socio-cultural, economic, and ecological challenges. They use various coping and adaptation strategies built on principles of flexibility and mobility. These strategies involves spreading and managing risks through communal utilization of land and water resources, diversification of herd composition, increased engagement in trade, wage-labour and cultivation (Little *etal.*, 2001). Attempts by the herders to surmount these challenges play a direct role in influencing the pastoral transformation process, as well as eventual changes in the risk management and livelihood of pastoral households.

In the Ethiopian peripheral arid and semi-arid areas, pastoralists operate in a context of social, institutional and environmental changes. These changes are mainly rooted in historical state-pastoralists relations that have disrupted the relations of local pastoralists with their environment and thereby their livelihoods and risk management strategies. State-led development interventions in the fertile Metehara plains of Awash Valley have been implemented at the cost of the environment and livelihood of the pastoralist community, which are dependent on the Awash River and its abundant pastures. Accordingly, the modernization programme of the previous regimes that depicted the Awash valley as nothing but agricultural land up for grab disrupted the environment where pastoralists had practised risk management and livelihood practices organized around water, pasture and livestock for centuries. Consequently, they now face drastic ecological changes in environments suffering from severe overgrazing, erosion, dwindling herds and subsequent famine and death.

With the coming to power of the Ethiopian people revolutionary democratic front (EPRDF), however, the political atmosphere and rhetoric changed, and various policies and strategies that permit the devolution of power to regions and local levels were formulated. Accordingly, unlike the previous pastophobic regimes, in post-1991 Ethiopia pastoralists have been given spaces in the political arena where they may have a say in the processes of decision-making that influence their livelihood. Various policies and programmes that directly or indirectly impact pastoralists have been formulated and put into practice. However, as the international hegemonic adaptation to impact of climate change discourse has taken centre stage in the national development policy of Ethiopia, pastoralists are also expected to progressively transform their livelihood with the help of projects and programmes that help them 'adapt to their environment'.

Accordingly, in the current discourses and practices of development in arid and semi-arid areas of Ethiopia, pastoralists' adaptation to climate change is pictured as necessary transformative processes that can be achieved through the means of technocratic government-driven projects to ameliorate the vulnerability of pastoralists to climate risk and help them achieve livelihood security. Though the government rhetorically recognizes the contribution of pastoralism to livestock production and has identified it as a driving force for further economic growth, the end goal is to radically transform this same livelihood into a settled way of life. This vision of the Ethiopian state is based on an all-encompassing discourse around the arid environment and the people inhabiting it. One line of the government discourse is that with the increased frequency of drought pastoral livelihoods are uniquely vulnerable. In this context, there have been efforts to improve environmental conservation and build community resilience as a means to manage risk and secure livelihoods. This premise which is founded on the environmental degradation discourse is used as a justification by the state to put in place projects meant to make the degraded arid environment 'hospitable' and modernize and transform the pastoral way of life. These government interventions have altered the society-environment nexus at the local level, which in turn differentially influences the livelihood security and risk management practices pursued on the ground. By situating the political ecology of risk management and livelihood security at the centre of Karrayu-environment-state interactions, this study examines the influence of government interventions on the Karrayu pastoralists' livelihoods and the environment, on the one hand, and the Karrayu pastoralists' resistance and accommodation to government interventions and how these shape the environment, on the other. The main research question is: how have state interventions and the associated environmental transformations been experienced and acted upon by the local pastoral communities in the arid and semi-arid Metehara plains of Upper Awash Valley, Ethiopia? More specifically, the study answers the following questions:

A. What are the historical trajectories of insecurity that have influenced Karrayu pastoralists' risk management and livelihood practices?
An understanding of local level pastoralists' risk management and livelihood security has to be located within broader historical trajectories that have influenced the interactions of pastoralists with their environment over the years and thus within their current risk management and livelihood strategies. The understanding of present-day risk management and livelihood security practices of the pastoralists firmly rests on an understanding of historical trajectories of resource politics that have shaped and influenced the interactions of the pastoralists with their environment. In answering this research question, this study explores the broader structural contexts in which livelihood insecurity arises for the pastoralists. Accordingly, it locates Karrayu pastoralists in the specific politi-cal ecology context that has evolved over recent decades.

B. How do Karrayu pastoralists continually practise livestock-based livelihood and risk management activities in the face of socio-environmental transformation?

This second question addresses the persistence of livestock-based pastoral risk management and livelihood practices under transformed socio-environmental conditions. Of particular interest here is the way the Karrayu pastoralists use their ability/agency to continually reorganize their livelihood and risk management practices in response to the changes in natural, social and political conditions, by activating or reactivating routine practices and traditional mechanisms for handling vulnerability. By emphasizing in particular the camel-based livestock production activities of the pastoralists, it will be shown how the Karrayuu pastoralists actively seek to maintain their livestock-based livelihoods in difficult Socio-environmental contexts.

C. How do Karrayu pastoralists take up and develop agro-pastoral livelihoods and risk management practices in the face of socio-environmental transformation?

The third research question focuses on the shift towards agro-pastoral livelihoods and risk management practices that the Karrayu have taken up in the face of changes in land tenure arrangements. The study presents the new resource arrangements surrounding the practice of farming and agro-pastoralism as a response to opportunities and constraints that the Karrayu are experiencing. By focusing on new mechanisms for managing scarce resources, I explore the new risk management and livelihood practices centred on agro-pastoral and farming activities. This helps us to understand that pastoral people do not just routinely respond to changing circumstances but actively use their agency and take part in processes that transform their livelihood practices.

D. How does research on context-specific risk management and livelihood security practices enable us to better reframe the mainstream perspective on adaptation?

The last specific research question relates to the theoretical and conceptual relevance of a political ecology approach, and puts local level risk management and livelihood practices in a broader context by comparing the empirical findings of the study with the hegemonic international perspective of adaptation in climate change research. In this regard, this study emphasises the importance of locating pastoralists' risk management and livelihood security practices in the context of the resource base on which they depend, and showing how they are related to these resources and how they interact with the state.

To answer these research questions, I decided to follow analytically and methodologically the political ecology approach which gives emphasis to pastoralists' modes of relations with natural resources and the active agency of pastoralists in these relationships that shapes their risk management and livelihood practices.

1.5 ORGANIZATION OF THE BOOK

This book consists of an introduction and seven chapters. This introductory chapter sets out the research rationale and background, states the point of departure of the study in the form of research questions, and places the dissertation within its conceptual and disciplinary field. Chapter Two provides the theoretical and conceptual basis by situating risk management and livelihood practices at the intersection of pastoralists-environment-state interactions. This is done using a political ecology approach that emphasizes pastoralists' agency in the face of constraining and enabling conditions. By bringing structuration theory into a political ecology approach that emphasizes the role of actors in shaping resource access and institutions, this chapter situates the work in the field of social geography. Building on this, Chapter Three introduces the research methodology and reflections on the procedure and practice of participatory and qualitative research. This chapter explains the methods I developed to enable me address my research objectives. It outlines the progression of my research, data collection instruments and analysis, followed by some ethical considerations.

Chapter Four presents the historical trajectories of livelihood insecurity as experienced by the Karrayu pastoralists over the past five to six decades. By locating the trajectories of livelihood insecurity at the intersection of pastoralists-environment-state interactions in the semi-arid Metehara plains, it sheds light on how the state's idea of national development was the driving force in the transformation of the landscape which resulted in changes in pastoralists-environment relations. These externally engineered changes were the root cause behind the disruptions in pastoralists' social and institutional setups that were initially founded on the logic of dealing with and managing seasonally variable natural resources. These changes in relations between pastoralists and natural resources have also been influenced by outside forces such as immigration of other pastoral or agro-pastoral groups into the Metehara plain, population pressure on the limited resources, and degradation of pasture land, to mention the major ones. This chapter traces the complex sources of risk and livelihood insecurity that have surfaced in Upper Awash valley over the past six to seven decades.

Following this, Chapter Five focuses on the discontinuity and reorganization of risk management and livelihood practices of the Karrayuu community in the face of changes in the natural resource base and local pastoralists' traditional mechanisms of handling vulnerability. Furthermore, this chapter delves into the forms of livestock-based livelihood and risk management practices adopted by the Karrayuu in the face of societal and environmental transformations. The Karrayuu pastoralists have used their agency to reorganize themselves around camel-based livestock herding and access distant resources that have been relatively less affected by the transformation of the landscape in the Metehara plain. Accordingly, this chapter explores the routine forms and the reorganized forms of livestock-based risk management and livelihood practices pursued by the Karrayu pastoralists.

In Chapter Six, emphasis is given to the risk management and livelihood practices of farmers and agro-pastoralists. This chapter gives particular emphasis to the way Karrayuu households organize themselves around the practices of agro-pastoralism and how they access the necessary resources to continue these activities. It locates the practice of agro-pastoralism at the intersection of the government's strategy of transforming the Karrayu way of life, internal differentiations within Karrayu social organization, and the immigration of the Ittu and associated changes in land use. The main intention of this chapter is to show how the interactions between outside forces of transformation and the pastoralists' own active agency lead to agro-pastoralism as a new form of risk management and livelihood practice.

In Chapter Seven, I consider the hegemonic adaptation discourse, that gives primacy to environmental stimuli, in connection with the previous empirical chapters that focus on processes of social, political and environmental interactions in shaping risk management and livelihood practices in arid and semi-arid areas of Ethiopia. In this chapter I argue that understanding the broader contexts in which risk management and livelihood practices take place helps to show that adaptation to climate risk is embedded in specific pastoralists-environment-state relations that have shaped the modes of relations around resource utilization. In doing so, this chapter tackles the last specific research question.

Chapter Eight concludes this book. In this chapter I summarize the major findings and contributions of my research by revisiting the questions that I formulated at the beginning. I also point out some directions for future research in the area of human-environment interactions.

2. ADAPTATION, VULNERABILITY AND LOCAL AGENCY: THEORETICAL AND CONCEPTUAL REFLECTIONS

2.1 INTRODUCTION

This chapter lays out the theoretical and conceptual frameworks that guide the research. It first starts by critically summarizing the conventional approach to adaptation research and the implication of this on our understanding of the sources of vulnerability. In doing this, I highlight the inadequacy of the conventional approach that puts high emphasis on climate-impacts in capturing the everyday social and political realities that shape vulnerability. It further explicates the two broader relationships of the concepts of adaptation and vulnerability; one that focuses on impacts and outcomes and the other relationship that focuses on contextual processes. Throughout this work I argue for the social embeddedness of vulnerability that emphasized on the differential ability of various actors in responding to change. In this regard, in order to better capture the processes of pastoralists-environment-state interactions in upper Awash valley and their implications on livelihood and risk management practices, I turn to political ecology approach that gives emphasis to interactions of various actors. Accordingly, risk management and livelihood security among the Karrayu pastoralists group of upper Awash valley are framed within broader and dynamic contexts of social, economic and political structures and institutions. Such framing helps us to locate processes of pastoralists-environment interactions and everyday risk management and livelihood practices within large-scale political-economic processes.

2.2 PUTTING THE CONCEPT OF ADAPTATION IN PERSPECTIVE

Although used in the 1980s, the popularity of the term adaptation has increased tremendously since it was first used in the United Nations Framework Convention on Climate Change (UNFCCC) in 1992 (UN, 1992). The year 1992 can be considered as a turning point for the wide application of this term connoting diverse meanings and interpretations within the climate change field. Such an increased application of the term, compounded with the taken-for-granted relevance that has been assigned to it in the fields of policy making and research, have definitely contributed to the increased conceptual complexity of adaptation. Accordingly, an intelligible and comprehensive adaptation theory becomes an illusion that has resulted in diverse understandings and disputations concerning its application in academic and policy circles (Schipper and Burton, 2009). Over the last two decades the approach to framing adaptation has exhibited progression from a focus on reducing impacts to vulnerability reduction. These two broad approaches to adaptation framing vary, depending on the emphasis that is given either to 'impacts' or 'vulnerability' in the analysis. In the major works in the field of climate change these two approaches are referred to as 'impact-led' and 'vulnerability-led' approaches to adaptation (Burton *et al.*, 2002; Smit and Pilifosova, 2003; Adger *et al.*, 2004; UNFCCC, 2005; Fussell and Klein, 2006; McGray *et al.*, 2007; Barnett, 2010).

Contrary to suggestions in respect of adaptation as a recipe for activities, the above two broad categorizations have allowed the expansion and evolution of critical understandings and interpretations of adaptation over time by applying theories from various disciplines. This has led to the recognition that different situations may require different responses. In other words, the consideration of whether climate change results in immediate damage or non-climatic stresses hinder responses to climate change will determine whether reducing impact or vulnerability is the most legitimate response. Accordingly, depending on the different articulation of meanings attached to these two opposing categories there are variations in terms of activities that can be considered as adaptation. These different kinds of activities are termed 'the adaptation space' (Ensor and Berger, 2009).

Linking adaptation to the concept of vulnerability and other interrelated concepts, such as resilience, has increased over time. This has resulted in the incorporation of theory from fields other than climate change that have utilized the concepts in a more refined manner. However, it should be clear that the changing understanding of the concept of adaptation has followed a complex path. In their efforts to come up with an "anatomy of adaptation", Smit and colleagues (2000) argue that the different approaches to adaptation are the result of considering three basic questions: Adaptation to what? Who or what adapts? How does adaptation occur? (Smit *et al.*, 2000: 223). On the basis of these three key questions, I summarize the core elements that differentiate impact-led from vulnerability-led approaches. Though I focus on the vulnerability-led approach throughout this chapter, I give a brief account of the conventional approach to adaptation and its pitfalls in the following sections.

2.3 THE CONVENTIONAL APPROACH TO ADAPTATION

The dominant impact-led approaches to adaptation in the climate change arena are known by various labels such as the "top-down" approach (Dessai, *et al.*, 2004), the "standard approach" (Burton *et al.*, 2002), and the "conventional approach" (Smit and Pilifosova, 2003). Despite the varied nomenclature, these perspectives share similar characteristics and have been widely applied in adaptation research during the 1990s (Burton *et al.*, 2002; Schipper, 2009). The continued prevalence of the impact-led approaches was also clear in the Fourth Assessment Report of the IPCC which based its method on this approach in the Working Group Two report (Carter *et al.*, 2007: 135). In general terms, due to their bold presence in the mitigation policy sphere, impact studies are seen as the first step for adaptation assessments. However, over the past two decades the apparent policy relevance and purpose of adaptation has evolved sharply. Impact analysis was designed to realize the ultimate objective of the UNFCCC, as partly described in Article 2, i.e. to prevent dangerous anthropogenic interference with the climate system (Burton *et al.*, 2002; Pittock and Jones, 2009), and thereby to formulate, implement and regularly update national and regional programmes containing measures to facilitate adequate adaptation to climate change, as clearly stated in Article 4.1 (b) (UN, 1992). An understanding of what constitutes a dangerous climate change was founded on the confirmation of its impacts. Moderating these 'dangerous' changes in climate and minimizing the impacts are the drivers of adaptation, parallel to mitigation measures. When assessing impacts was first started during the 1990s, adaptation was viewed as an alternative to mitigation. This dominant view during that time helped to subdue perspectives that put high emphasis on minimizing greenhouse gas emissions (Pielke, 1998; Smit and Pilifosova, 2001; Burton *et al.*, 2002; Klein *et al.*, 2003; Schipper, 2009). The primary objective of analysing the impact of climate change during this time was to decide on how to maintain the ideal balance between adaptation and mitigation measures and thereby feed into the mitigation policy. Accordingly, the impact-led approach is basically occupied in answering questions revolving around issues such as the 'extent of the climate change problem'; comparison of the 'costs of climate change' with that of the 'costs of greenhouse gas mitigation' (O'Brien *et al.*, 2004: 3). The influence of prioritizing such issues has led to the consideration of mitigation, adaptation and impacts by the same working group. This has clearly been reproduced in the Second Assessment Report of the inter-governmental panel on climate change. In a nutshell, the primary focus of the impact-led approach to climate change was on the costs of damage arising from climate change (Parry and Carter, 1998). Following this line, adaptation was seen as a specific measure designed to address the impacts. The principal objective of the impact-led approach is to assess the damage-costs of climate change and the alteration adaptation measures may bring (Smit and Wandel, 2006). Despite the rise of adaptation to a high position on the policy agenda as a necessary activity regardless of mitigation activities, impact-led perspectives still dominate the understanding of adaptation. Some scholars associate the dominance and prevalence of impact-led approaches in adaptation with the international climate policy agenda that is essentially mitigation-oriented (Schipper, 2009). For instance, UNFCCC clearly stipulates that the reduction of greenhouse gas emissions is at the top of the agenda and that there is no room for adaptation. Scholars further argue that within the current framework of the UNFCCC addressing adaptation is challenging (Pielke, 2005; Burton, 2009).

2.4 THE INADEQUACY OF THE CONVENTIONAL APPROACH

In developing countries, the impact-led approach lags behind in facilitating concrete adaptation outcomes, as predicted in Article 4(4) of the UNFCCC. In the late 1990s and early 2000s, recognition of the inadequacy of mitigation measures has paved the way for increased attention to adaptation as a vital complement rather than an alternative to mitigation. Driven largely by the concerns of developing countries, adaptation started to receive attention as a separate policy agenda independent of the mitigation programme within the climate change arena (Huq *et al.*, 2003; Sokona and Huq, 2002; Huq and Reid, 2004; Schipper, 2009). Accordingly, adaptation as an agenda appears for the first time in the IPCC under Working Group Two. With an increased emphasis on adaptation within the policy circle, a growing concern for vulnerability can also be noticed (Schipper, 2009). International negotiations, mainly led by developing countries, facilitated the development of vulnerability-led approaches to adaptation. This increased the urgency to incorporate relevant provisions for adaptation and adaptation policy that resulted in the Marrakesh Accords (UNFCCC, 2002). This agreement clearly stipulates the support of adaptation activities in developing countries through the supply of funding under the UNFCCC and its Kyoto protocol (Burton *et al.*, 2002; Schipper, 2009). The Marrakesh Accords had important implications for the increased role of the concepts of development and vulnerability in adaptation research within the climate change policy circle by pointing out the specific challenges that developing countries are experiencing (Adger *et al.*, 2003). Consequently, studies on adaptation shifted their focus to the procedure of adaptation and allocation of funding for selected adaptation activities. This is also reflected in Article 4.4 which stipulates increased assistance by developed countries to support vulnerable developing countries and help them to curb the effects of climate change by means of financial assistance for adaptation measures (UN, 1992). The greater priorities that had been given to an impact-centred approach to adaptation gave way to a focus on a vulnerability-centred perspective.

Accordingly, there is a complete shift in the objective and expected outcome of adaptation. New premises were required to address a different conceptual task through the lens of the vulnerability-led perspective. O'Brien and colleagues (2004) pinpointed the challenge of identifying who is vulnerable to climate change and the reasons for this, as well as the mechanisms of reducing vulnerabilities in relation to climate change as the major concerns of the vulnerability-led approach. This has steered the central concerns away from determining the extent to which adaptation reduces the need for mitigation and towards finding out the priority for adaptation and formulating policies by consulting various stakeholders (Burton *et al.*, 2002; Carter *et al.*, 2007). The progression towards a vulnerability-led approach demanded consideration of the broader socio-political, cultural and institutional contexts within which adaptation to climate change takes place. Following this line, Adger and colleagues (2003) contend for reflection on broader social and environmental contexts that influence the ways in which local people react to and cope with climate hazards. Hence, in contrast to the impact-led approach in adaptation research, the vulnerability-led approach puts emphasis on broader structural forces that adversely affect and constrain the smooth functioning of particular systems in the face of climatic stimuli (Kelly and Adger, 2000; O'Brien *et al.*, 2004; Adger *et al.*, 2004; UNFCCC, 2005; Smit and Wandel, 2006; Turner *et al.*, 2003). The consideration of the social context within which local people cope with and adapt to climate hazards has prompted scholars to label the vulnerability-led approach as the 'bottom-up' approach (Dessai *et al.*, 2004), or the 'second generation' of adaptation studies (UNFCCC, 2005; Fussel and Klein, 2006).

2.5 THE CONCEPT OF VULNERABILITY IN ADAPTATION RESEARCH

In the two broad approaches to adaptation research explained above, the concept of vulnerability has different roles in relation to adaptation. While impact-led approaches generally begin assessment with scenarios of long-term average changes, and focus on 'specific adaptations' to reduce future potential impacts, vulnerability-led approaches begin with stresses and the contextual reasons for these stresses (Kelly and Adger, 2000; Smit and Pilifosova, 2003; Smit and Wandel, 2006). The purpose of this section is to analyse the concept of vulnerability and its variable relationship to the concept of adaptation.

The concept of vulnerability in the field of climate change is surrounded by vagueness (Brooks, 2003). There are various understandings of vulnerability which can be categorized in different domains (Fussel, 2007). Climate change research pulls together scholars from various disciplines such as disaster management, food security, and development studies (Brooks, 2003; Fussel, 2007). The assumptions underwriting different interpretations of vulnerability vary among research fields and scholarly communities and over time. The variation in the interpretation of vulnerability is mainly attributed to variations in the epistemological and methodological orientations of the different fields involved in vulnerability research (Cutter, 1996). In the context of local-level risk management and livelihood practices, a shared conceptual understanding of vulnerability within disaster relief, development and climate change research is recommended (O'Brien *et al.*, 2006). Conversely, scholars in these fields follow their own line of thought on the concept of vulnerability and cooperate minimally to come up with a shared understanding. The problematization and framing of research, as well as policy orientations with respect to adaptation and vulnerability, have also been seriously influenced by a lack of shared understanding among scholars (O'Brien *et al.*, 2004). The connection between the two concepts is usually underestimated (Schipper 2007; 2009). There are two crucial ways in which they can be related that form the corner-stones for the different meanings attached to them (Schipper, 2007; Schipper and Burton, 2009). The first one sees vulnerability reduction as a facilitator of adaptation to climate change, while the second assumes that adaptation to climate change impacts can reduce vulnerability.

The second view of the relationship between the two concepts suggests that climate stimuli are part and parcel of vulnerability, and minimizing the impacts of climate stimuli will minimize vulnerability to climate change. This line of reasoning puts climate change at the centre of the problem. On the other hand, a view of the relationship between the two concepts as one in which vulnerability reduction facilitates adaptation to climate change impacts suggests a far broader interpretation of adaptation and vulnerability. This view of the relationship between the two concepts implies that vulnerability to climate change is related to broader environmental and social situations than mere climate change impacts. Hence, vulnerability reduction catalyses adaptations to the impacts of climate change through development activities under the particular social conditions. According to this interpretation, the underlying structural forces are the driving forces behind people's reaction to climate change impacts. Hence, the concept of adaptation means something different in each of the two contending interpretations. This differential interpretation of vulnerability in relation to adaptation matches 'end-point' (adaptation to climate change impacts can reduce vulnerability) and 'starting-point' vulnerability (vulnerability reduction as a facilitator of adaptation to climate change) (Kelly and Adger, 2000; O'Brien *et al.*, 2004; Fussell and Klein, 2006; Fussell, 2007; Ensor and Berger, 2009). The key conceptual differences are outlined in the following figure.

Figure 1: Interpretation of vulnerability in climate change adaptation research

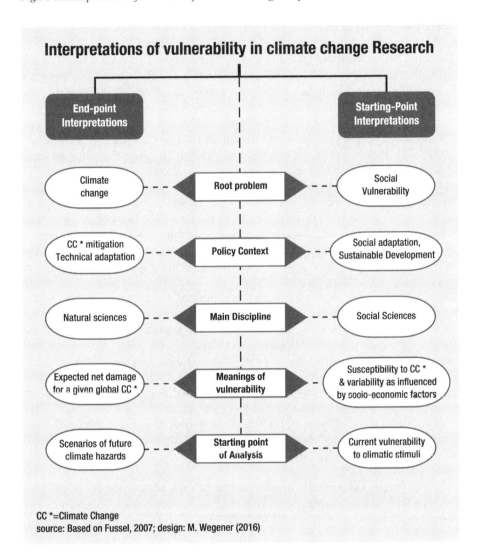

2.5.1 Socially embedded vulnerability

The central premise in a starting-point analysis of vulnerability rests on an understanding of non-climatic factors and social processes that constrain people's adaptation and coping mechanisms in respect of climate change. Furthermore, since vulnerability exists even in the absence of particular physical hazards, the broader socio-economic conditions that hinder or facilitate people's strategies should be the centre of analysis in starting-point vulnerability. So, vulnerability happens at the "intersection of external risks and stresses and the internal lack of peoples' capacity to respond to and cope with damaging loss" (Chambers, 1989: 1). Unlike the very mechanistic viewpoint of end-point interpretation, it is difficult to determine the nature of vulnerability and the mechanisms to reduce it based on the particular type of threat that people are vulnerable to. This shows that people's vulnerability to specific threats is located in social conditions. With respect to the starting-point conceptualization of vulnerability, Kelly and Adger contend that "social vulnerability is the capacity of individuals and social groups to... cope with, or adapt to any external stress placed on their livelihoods and well-being" (2000: 325). In this regard, the reduction of vulnerability should involve an array of activities that improve groups' or individuals' capacity to deal with current or future climate-related stresses. Vulnerability can be expressed in various ways depending on the types of external threat. However, the underlying social dynamics that create vulnerability have to be the central concern of starting-point vulnerability (Allen, 2003). In other words, despite the various sources of external threats, vulnerability is mediated by social factors. Kelly and Adger (2000) use the metaphor of the "wounded soldier" in order to better explicate the concept of vulnerability by construing the wounds as constraints that make it hard to effectively deal with external stresses regardless of the particular nature of the external stressors. This broad perspective, which sometimes is also called the 'double exposure' framework, sees vulnerability to climate change as an outcome of the interaction of non-climatic social factors with external stresses that constrain the locals' ability to respond where climate change is one of them (Turner *et al.*, 2003; Reid and Vogel, 2006; Yohe *et al.*, 2007; Silva *et al.*, 2010). In other words, local landscapes of vulnerability are the result of a combination of broader social conditions and external processes of environmental change (Silva *et al.*, 2010). Thus, climate stress comes on top of the social fabric of local landscapes and tells a lot about other pressing social stresses, such as restricted access to land, which have already been influencing the livelihoods of the people (Reid and Vogel, 2006).

Accordingly, vulnerability to climate change does not happen out of the blue and function independently from vulnerability to other social stressors. The starting-point approach opposes the identification of any particular major stressor. Rather the starting-point perspective argues that it is not always appropriate to single out the impact of a specific physical hazard to define the nature of vulnerability, as vulnerability is a combined outcome of 'non-climate' processes. In the starting-point perspective of vulnerability, social fabrics are at the centre of the analysis, with a prime focus on peoples' ability to adapt (Adger and Kelly, 1999; Kelly and Adger, 2000; O'Brien *et al.*, 2004; Fussell, 2007). This line of interpretation is clearly social-science oriented and strives to disentangle the social conditions that affect the differential distribution of vulnerability among society's constituent members. In short, the starting-point interpretation of vulnerability is based on an understanding of the relationship between adaptation and vulnerability as the one in which vulnerability reduction enables adaptation.

Accordingly the focus of adaptation is on social solutions by putting emphasis on peoples' capability to adapt (Eriksen and Kelly, 2007).

Scholars also contend that the starting-point interpretation of vulnerability has implications for strategies that are less related to climate stress. Rather, these strategies are integral to activities at a local level and tailored towards better livelihoods, poverty reduction and so forth (Smit and Wandel 2006; Schipper, 2007; McGray *et al.*, 2007). In this regard, increased concern over climate change accelerates efforts in development practice to reduce vulnerability (Wilbanks, 2003; Schipper, 2007). This clearly shows that theoretically there is not much difference between sustainable development processes and vulnerability-led adaptation to climate change (Davidson *et al.*, 2003).

2.5.2 Not always passive victims: focus on the locals' agency

Unlike end-point vulnerability, which paints a picture of the locals as passive victims exposed to threats and hazards, the starting-point interpretation acknowledges the ability of those affected to respond to climate change through their active engagement (Hewitt, 1983b; O'Keefe *et al.*, 1976; Wisner *et al.*, 2004). The recognition of active agents in the face of hazards also has a different implication for the framing of actions. In this case the actions that arise out of starting-point interpretations revolve around helping to strengthen people's own ability to react, and thereby adapt, rather than devising mechanisms of protection which are at the centre of end-point vulnerability (Eriksen and Kelly, 2007). Many social scientists (Brooks, 2003; Eriksen *et al.*, 2005) frame their understanding of vulnerability by focusing on people's potential to deal with stress. Analysis based on the starting-point approach to vulnerability also helps us to appreciate how people's strategies for handling climate-related extreme events have evolved over the years (Yamin, 2004; Heijmans, 2004). Thus, looking at vulnerability through the lens of starting-point analysis also tones down exaggerated, fatalistic views of the impact of climate change on the local people. As discussed above, starting-point interpretations of vulnerability within the climate change adaptation sphere bring attention to the ability of people to respond to climate change by focusing on adaptive capacity and the related concept of resilience. This is shaped primarily by social, rather than biophysical or climatic processes operating in specific contexts. Focusing on the response capacity of people therefore emphasizes questions about why some groups may have more or less capacity to respond than others, or what factors and processes facilitate and constrain the capacity to adapt. Within a starting-point interpretation the answers are in the Socio-economic and political structures that shape differential access to resources which secure livelihoods and therefore shape adaptation. A key research question within this field is: "what political and economic arrangements accelerate or decelerate reductions and enhancements in human vulnerability …?" (Turner and Robbins, 2008: 300). Analysis adhering to this conceptual approach therefore aims to be 'explanatory' rather than 'descriptive' as in an end-point framework (Fussell, 2007).

In line with the starting point analysis of vulnerability that gives emphasis to the social aspect of vulnerability, political ecologists open analyses to the diverse range of factors that influence the knowledge and actions of rural producers in their attempt to access and use resources, from the conception of power to agents at various levels, while appreciating the social nature of agency. Human agency is the capacity of human beings to make and exercise choices within their own cultural milieu. Emirbayer and Mische (1998: 970) define agency as "the temporally constructed engagement by actors of different structural environments...which, through the interplay of habit, imagination, and judgement, both reproduces and transforms those structures in interactive response to the problems posed by changing historical situations". In their extended analysis of the concept of agency, Emirbayer and Mische (1998) further explore the three analytical dimensions of agency that I have also found relevant for the understanding of pastoralists' agency within socio-environmental structures.

The first element of agency – the iterational aspect – refers to the actors' inclination towards reviving past patterns of thinking and doing things and routinely making them part and parcel of practical activities which help them to sustain their institutions, identities and interactions over time. For instance, in the case of the Karrayu community, despite the major structural constraints in the form of 'development interventions' that hamper their pastoral way of life, they continue to sustain their routine pastoral practices and hence maintain their identity. However, this does not mean that everything is intact and working well within the Karrayu pastoral community. The Karrayu pastoralists also evaluate and respond to new demands emerging from their social and environmental conditions. This leads us to the second analytical aspect – the practical-evaluative element of agency. Through this agency, actors make practical and normative judgements to choose among various pathways of action in the light of developments in their surroundings. The third element of agency, which is also relevant for this study, is the projective aspect of agency. Emirbayer and Mische (1998) suggest that projective agency is "the imaginative generation of possible future trajectories of action". These trajectories are usually outcomes of the actors' reconfiguration of sedimented thoughts and actions based on their aspirations and anxiety for the future. All these three elements may appear in tandem with one another at a given time under the influence of broader structural conditions. Based on the options and resources that they have created with and through their agency, the actors try to influence the structural conditions in order to persue their objectives.

The emphasis on active agents and local people's potential is further elaborated in the disaster studies literature through the concept of 'capacity' (Davis, 2004; Wisner *et al.*, 2004). This emphasis on people's ability helps us to understand the connection between resources and assets at people's disposal, on the one hand, and their agency to access and utilize these resources on the other, which is necessary to be able to respond to hazards (Gaillard, 2010). Thus, many of the factors that reduce people's potential agency are situated in everyday interactions between active agents and their environment.

2.6 POLITICAL ECOLOGY AND THE VULNERABILITY PARADIGM

Clear use of the term 'political ecology' started with the seminal work of Blaikie and Brookfield (1987) (Turner and Robbins, 2008). The work of Forsyth (2003) is based on a broad definition of political ecology that takes into account general contexts within which both environmental problems and management issues arise. Many authors (such as Bohle *et al.*, 1994) have also defined political ecology as a combination of political economy and human ecology approaches to the interactions between nature and society. Political ecology varies from the political economy approach since it puts emphasis on the environment as an independent variable that structures social relations. In other words, the influence of the environment on societal change is recognized in the political ecology approach. Contrary to this, the political economy approach focuses on the impact of economic structures in influencing the environment. Accordingly, little consideration is given to the physical environment (Greenberg and Park, 1994). At an initial stage this served as a drive for the augmentation of the broadly defined political ecology tradition (Greenberg and Park, 1994; McLaughlin and Dietz, 2008). Unlike the human ecology perspective, political ecology puts the structural problems arising from economic and political ideologies at the centre of environmental problems and suggests alteration of these structural forces as a necessary step towards tackling environmental problems. Consideration of the causes of environmental problems located in social, political and economic contexts (Watts, 1983) was the basis on which structuralists Watts (1983) and Hewitt (1983a) criticized the works of behaviouralists Burton, Kates and White (1978). This helped to create a new epistemological position in contrast to earlier works of human ecology (Watts, 1983). In this regard, the concept of vulnerability perfectly fits into political ecology as it reveals how nature and culture are mutually connected and constituted (Oliver-Smith, 2004). One of the ways in which this mutual constituency is clearly manifested is the occurrence of disasters. For instance, the livelihoods of communities in the Sahel are vulnerable to climate change mainly due to ill-advised policies geared towards increased economic growth at a national level. Here one cannot deny the impact of climate change as manifested in recurrent droughts. However, development policies "dictated by global economic paradigms" sharply minimize the capacity of the locals to deal with drought and hence render them vulnerable (Adger and Brooks, 2003: 29).

Therefore, scholars contend that the political ecology approach needs to locate social interactions within the broader framework of ecological processes and should not be trapped in the futile debate of nature/culture duality (Gezon, 2006; Latour, 2004; Zimmerer and Bassett, 2003). In their emphasis on the decisive and influential role of the bio-physical environment in influencing human-environment relations, Zimmerer and Bassett further argue that "the environment is not simply a stage or arena in which struggles over resource access and control take place" (2003: 3). So, political ecology has to indicate not only how the environment is an outcome of persistent competition over material and tangible activities, but also needs to focus on competition over meanings (Bryant, 1998: 82). The overarching point here is that actors actively manipulate symbolic meaning to their own ends by outlining their interests and strategies within particular cultural frames. In the context of vulnerability, divergent frames produce varying definitions of vulnerability in terms of its character and causal structure. What causes vulnerability and disaster, and therefore what actions are needed to respond to these, is culturally constructed. McLaughlin and Dietz (2008) observe that such frames form the basis for coordinating action in respect of a problem, and thus can represent struggles for domination in how the problem – and its solution – is perceived and acted upon. Thus, coordinated action, such as disaster risk reduction and livelihood security, are value-laden and can represent struggles for legitimacy and power among different actors. Locally-based institutions derive their frames from cultural sets of meanings within local villages, which contrast considerably with the perceptions and understandings that prevail on an international scale.

Whether vulnerability is a socially constructed issue or not is highly contested. At one extreme we find the 'radical constructivists' who argue that the biophysical environment itself is an outcome of people's perception and there is no independent interconnection between vulnerability and disaster. What exists is only human perception of these events (Oliver-Smith, 2004; Bankoff, 2004; McLaughlin and Dietz, 2008). Scholars in this group further contend that risk is not inherent and hazards are not realistic phenomena. Rather, they are constitutive of social natures where hazards and risks are culturally bounded (Bankoff, 2001). They also criticize the realists for their separation of nature from culture and for handling nature independently of human intervention. However, the radical constructivists have been criticized for reducing natural hazards to something that is discussed, ignoring the autonomy of the dynamic natural world by reducing it to a social construction.

On the other hand, moderate social constructivists acknowledge the reality of an external world, but accept that beliefs about that world are imperfect (McLaughlin and Dietz, 2008). They understand the biophysical environment as an independent causal force in vulnerability, but accept that vulnerability is socially embedded, and that responses to vulnerability are culturally bounded (Wisner *et al.*, 2004; Wisner *et al.*, 2004; Bankoff, 2001). This approach is also called critical realism and is the basis of a critical political ecology (Escobar, 1999; Forsyth, 2003). According to this approach, there is an objective biophysical nature, but this is always mediated through people's beliefs and their understandings of this biophysical nature. According to this perspective, biophysical phenomena are not inherently hazardous. A biophysical phenomenon is a hazard if it has damaging consequences that destabilize people's livelihoods. Therefore, hazard implies that there is a risk, or at least the possibility of the incidence of a disastrous episode.

In their pressure-and-release model of disaster, Wisner and colleagues (2004) point out that risk is constituted by the interplay between certain non-human processes or events and underlying social dynamics and unsafe conditions, such as those produced by market-oriented development practices that separate the local people from their resource base. In this sense, risk and hazard are themselves forms of social mediation, but they are simultaneously reliant on bio-physical agency (Tsing, 2005). Furthermore, although risk and hazard involve perception which is subject to cultural intercession, that intercession does not refute the essential biophysical reality of that particular event.

Despite their variations in the conception of vulnerability, constructivists add to debates in the field of disaster risk studies. They introduced the cultural aspect in disaster studies and increased our understanding of the connection between people's values and their vulnerability. They have helped us locate vulnerability in its particular special and historical contexts. They have contributed to disaster studies by focusing on the capacity of the local population to respond effectively in the face of disasters (McLaughlin and Dietz, 2008). They are also critical of associating natural hazards with societal destabilization and malfunctioning in the inner workings of everyday societal activities (Oliver-Smith, 2004; Wisner *et al.*, 2004). As shown in this study, Karrayu pastoralists had integrated drought hazards into the regular workings of their community and did not consider these to be outside the normal order of things. However, due to changes in their modes of relationship with their natural resource base, their framing of risk management and livelihood practices has changed over time. Through an understanding of the interactions between societies and environments within the broader political and economic contexts, political ecology elucidates the vulnerability that arises from the whole constellation (Bohle *et al.*, 1994; Oliver-Smith, 2004; Watts and Bohle, 1993; Wisner *et al.*, 2004).

2.7 LOCATING ACTORS IN POLITICAL ECOLOGY

Since its inception, political ecology scholarship has advanced the analysis of the interactions between social and environmental change. Issues of social and environmental justice are central in political ecology, which aims not only to explore dimensions of power and marginalisation in processes of environmental change but also the "alternatives, adaptations and creative human action" (Robbins 2012:20) in the face of such changes. In sum, political ecology offers a critical lens for exploring the complex and multifaceted issues of risk management and livelihood security in arid and semi-arid environments, embedded in broader contexts of social, economic and political structures and institutions. Thus it also provides an eclectic and important frame for linking local level processes and everyday struggles to large-scale political-economic processes and policy narratives.

In this book, in line with Paulson and Gezon (2005), I take a political ecology approach that focuses on the interrelations and practices of people as they engage themselves in productive activities. Such a focus on the material aspect of a material world makes a political ecology approach totally political (Biersack and Greenberg, 2006: 28). In this sense, power is continually implicated and emergent in the inner workings of material social processes, material practices, and the outcomes of material struggles (Roseberry, 1989).

Figure 2: Local actors in a broader political-ecology context

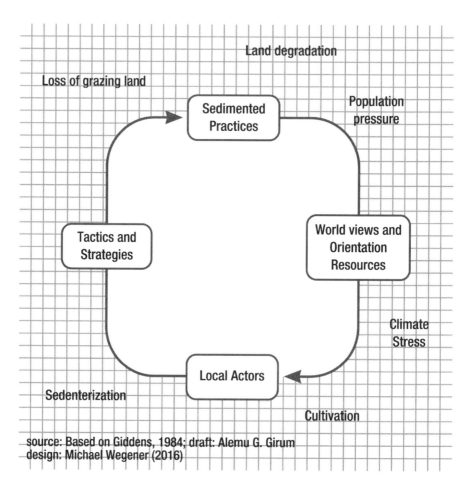

source: Based on Giddens, 1984; draft: Alemu G. Girum
design: Michael Wegener (2016)

As this study illustrates, inequalities and politics are part and parcel of all human-environment interactions that influence access to resources and resource use practices (Gezon, 2006) and it employs the political ecology approach that emphasizes on power relations in all spheres of social life (Paulson and Gezon, 2005).

Political ecology strives not only to capture how individuals and communities are affected by environmental change but also to explicate people's agency and creativity in the ways such problems are tackled, negotiated and contested (Rocheleau, *et al.* 1996). Paul Robbins refers to these twin goals as 'the hatchet' and 'the seed' of political ecology (2012:99). A focus on agency underlines that social structures, rules and norms are not casted on stone; while some actions and practices serve to reproduce and strengthen those structures, others serve to re-interpret and change them (Giddens, 1984; Brown and Westaway, 2011). As argued by Benedict Kerkvilet (2009), the latter form of actions and practices often takes place in 'in-between' or 'back-door' spaces that are not always evident. Such forms of 'everyday politics' may contain strategies that people formulate in order to generate 'room for manoeuvre' in struggles over resources or to seek out alternate development pathways. In line with Ann Swidler, strategies can thereby be understood as "persistent ways of ordering action over time" (1986:273).

Such an understanding sees power as residing in both individual actors and structures which are always subject to challenge and negotiation, as well as coercion and submission. In line with this, Giddens defines power as "the transformative capacity of human action" (1976:110), and the "capability of an actor to achieve his or her will" (1979:69). Power can also be a characteristic of groups with resources as its structural foundation (Giddens, 1979). Within the pastoral system, possession of a large number of animals increases the power of individuals and groups, but this power is diffused through the lateral reciprocity of resource exchange. For instance, those Karrayu households who have managed to adapt better to the existing conditions have a better livelihood and become better off. However, through social obligations they share their resources with others who are in a disadvantaged position within the community.

Accordingly, I regard the Karrayuu as agents in their own right. In line with this, I argue that individual Karrayu do not respond aimlessly to structural changes emanating from above. This approach emphasizes the dynamic engagement of individual actors in the processes of negotiation and production of social change, while at the same time pursuing their own benefits. Such an interaction between actors and the social structure is captured by Giddens' explication of the 'duality' of agents and structure. Giddens focuses on the 'duality' between structure and agency, each interrelated with and effecting change in the other. Giddens' idea of duality of structure explicates why human agency produces social structures which serve as the platform for their own replication, again with the help of actor's agency and activities (Giddens, 1976:121). Yet structure is at the same time the platform in which societal rules are modified, and in this way all societies are dynamic (Giddens, 1979:18). Giddens refers to "sequences of change which are of medium term duration, but which have far reaching consequences for the society or region in question" (Giddens, 1979:228). The broader socio-political and ecological trajectories of vulnerability, presented in Chapter Four, produce and reproduce the present pastoralists' risk management and livelihood practices which in turn impacts the broader structural context through the locals' appropriation, resistance and adaptation strategies.

According to Giddens (1979:64) structure refers to "structuring properties... the 'binding' of time and space in social systems ... understood as the rules and resources, recursively implicated in the reproduction of social systems." They include knowledge on the part of the actors of what must be done. This knowledge may be "practical" and tacit, something an actor may not be able to explain, or it may be "discursive", something the actor is able to discuss and analyse. The production of practices mobilized by actors through stocks of knowledge and capabilities necessitates access to resources under the constraint of rules (Giddens, 1976:108). It is clear that one actor alone cannot replicate a social system; social structure implies interaction between actors. Actors use their own stocks of knowledge, their capabilities, and their access to resources, and are constrained by rules, thus recreating the structure that those rules and resources constitute (Giddens, 1979:71). The actors reflect upon the intended and unintended consequences of their practices and interactions, adjusting their reactions and thus modifying the structure. One outcome of actors' interaction in a semi-arid environment could be vulnerability which is embedded in complex social relations and processes (Nelson and Finan, 2009:305).

In the process of social interaction, where power is continuously at play, some meanings are advantaged over others. To put it differently, this is crucial not only because meanings shape the way specific actions are grasped and outlined but also because certain meanings, such as those expressed in norms and rules which have important effects on how pastoralists access resources and use them, become manifested in material ways that shape localized adaptation and vulnerabilities in the face of social and climatic changes.

2.8 SUMMARY

In this theoretical chapter, by adding a consideration of human agency to the political ecology perspective, I have argued that the local peoples' risk management and livelihood practices need to be considered within broader historical trajectories of development that disrupt human-environment interactions, rather than as mere adaptation responses to climate change impacts.

Accordingly, I have outlined the mainstream international adaptation discourse, showing the progression from a focus on impacts to a focus on vulnerability. I have unpacked the concept of vulnerability and in particular, examined its theoretical roots in disaster risk reduction scholarship. The unpacking of the concept of vulnerability helps us to better capture political and institutional processes that mediate access to resources and utilization, which in turn influence how risk management and livelihood practices are being pursued on the ground. I further argue that managing risk and securing livelihoods is highly dependent on socio-institutional factors. Throughout this book I argue that in order to better understand the risk management and livelihood practices of local communities, we need to situate these practices in broader historical trajectories of pastoralists-environment-state interactions that have shaped and influenced contemporary livelihood activities of the karrayu community.

3. RESEARCHING WITH THE LOCALS: METHODOLOGICAL REFLECTIONS

3.1 INTRODUCTION

This chapter outlines the research approach used in this work. In particular, it discusses the methods I developed to enable me to address my main research question: how have state interventions and the associated environmental transformations been experienced and acted upon by the local pastoral communities in the arid and semi-arid Metehara plains of Upper Awash Valley, Ethiopia? To answer this question, it was necessary to find a method that would enable me to capture local constructs of vulnerability in the face of social change and climate stress.

In Chapter Two it is shown that the power differentials of various actors in the community affect their vulnerability to both social and climatic changes within the sphere in which pastoralists operate. When doing research in a resource-dependent community like that of the Karrayu pastoralists in the arid and semi-arid areas of Upper Awash Valley, it is important to consider that methodologies must be closely linked to the specific issues and questions the study is attempting to address. These issues are critically embedded in the historical undercurrents of the people and the specific geographical context, and they carry inferences for the research. Thus, there is a need to consider and appreciate the context within which such issues emerge and continue. Another purpose of this chapter is to critically reflect upon how I have undertaken my research and in doing so, to examine established concepts of vulnerability in the climate change tradition.

I begin the chapter by introducing the field site of Fentalle *Woreda* in Upper Awash Valley. I then outline the broad methodological approach, which is based on qualitative and participatory approaches in geography. The research process is discussed, with details of how the study was set up and how it progressed through a number of key activities in the field. Following this, I discuss the specific research methods used for data collection, writeup and analysis, and some ethical considerations that emerged during the research process.

3.2 UPPER AWASH VALLEY, FENTALLE *WOREDA*: DESCRIPTIONS

3.2.1 Research area and the people

The Karrayu pastoral communities inhabit the Metehara plain in Upper Awash Valley. According to Muderis (1998:54) the Karrayu have lived in the area since the 16th century. The Karrayu pastoralists are divided into Dullacha and Baso moieties and further sub-moiety structures. The Karrayu land is situated in Fentalle District in the Oromiya region of eastern Ethiopia. The main town of the district, Metehara, is located some 200 km to the east of Addis Ababa. The district covers the Metehara plain and the environs of Mount Fentalle on the fringes of East Shawa. According to the 2007 census, Fentalle *Woreda* is home for 82,225 people (53.9% male and 46.1% female). The Addis Ababa-Djibouti highway and railway which bind Ethiopia to the outside world pass across the district. Geographically, the Fentalle district is located between 30° 30′ –40° 11` longitude east and 8° 42` –9° latitude north. The altitude of the district ranges from 980 metres on the plains to 2007 metres at the peak of Mount Fentalle. The area is covered with scanty bushes and Acacia trees. The total land area of the *Woreda* is estimated at 133,963.66 hectares. The *Woreda* Pastoral Development Office divides this into arable land (19,611 ha), land reclaimed with grass and other vegetation (457 ha), and others (swampy, water-covered, urban lands and rugged topography (109,897 ha). The *Woreda* has rich livestock resources – 65,412 cattle; 226,628 goats; 346,695 sheep; 120,958 camels; 13,064 equines (12,714 donkeys and 350 male horses); and 8,342 chickens. The Karrayu raise cattle, camels and small ruminants, collectively known as *karra sadeen* – the three stocks – for food, exchange, ritual values, marriage transactions, maintenance of survival networks consisting of resource redistribution and mutual support among clan members and the neighbourhood.

Figure 3: Map of the study area showing Fentalle Woreda

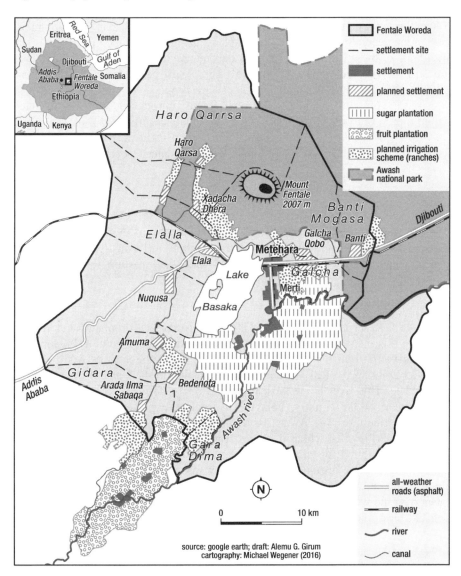

The *Woreda* is bordered by Afar Region in the north, Amhara Region in the northwest, Boset *Woreda* (East Shawa Zone) in the west, Merti *Woreda* (Arsi Zone) in the south and West Hararghe Zone in the east. According to the *Woreda* Pastoral Development office, there are eighteen rural and two urban *Kebeles* of Metahara town. The *Woreda* Rural Land and Environmental Protection Office classifies the rural *Kebeles* into three categories: 'pure pastoralists', 'agro-pastoralists' and 'farmers'. Most of the 'farmers' are people who were displaced and resettled when they lost their rangeland to Metahara Sugar Cane Plantation (MSF) during Haile Selassie's time (for detailed explanation, see Chapter Four). Since then, they have been tilling the land as semi-sedentary communities, cultivating cereals and some vegetables, using traditional irrigation. The livelihood of these communities largely depends on livestock production and they are still active practitioners of mobile pastoralism, particularly camels. Those falling under the 'agro-pastoralist' category are people living in those *Kebeles* in which the Oromia National Regional State introduced modern irrigated agriculture in the year 2010. The *Kebeles* are inhabited primarily by Karrayu and Itu Oromos, while Argoba farmers and some Somali segments have expanded into Tututi and Ilala from Amhara and Somali Regional states respectively.

3.2.2 Physical features

The physical features of the Karrayu territory consist of the Beseka Lake, the Kessem River, Mount Fentalle, and the Awash River with extensive lowland plains, which are an important source of water and pasture for the Karrayu livestock.

The Awash River
The Awash River is one of the few Ethiopian rivers that originates and ends in Ethiopia. It is the second largest river, covering a distance of some 1,200 km. The Awash sets off from the highlands around Ginchi, to the west of Addis Ababa, and after crossing the Rift Valley it finally vanishes in the Danakil Depression in Lake Abbe. This river is the most important one in terms of contributing to the development of the country, both for hydroelectric power generation and for irrigated agriculture. The Awash River drains the Metehara Plains and enriches them with fertile silts, which it carries from the highlands. This has provided a source of livelihood for pastoralists such as the Karrayu for centuries. But now, irrigated agriculture occupies the former dry season grazing areas on the Awash flood plains at Abadir and Metahara/Merti. And the Awash National Park has taken over the wet season pastures. Because of these schemes, the pastoralists are denied access to the waters of the Awash River.

Lake Beseka
This lake is located south of Fentalle Mountain and north of Metehara Sugar Factory on the outskirts of Metehara Town. Within the last forty years, since the beginning of the irrigation scheme, the surface area of the lake has increased from 3.3 km^2 to 35 km^2. The increase of the surface of the lake is one of the factors that have minimized access to pasture land for the Karrayu. Moreover, the lake cannot be used for both human and livestock water consumption because of its high salt and fluorine content. It only provides fish which is served in the hotels of Metehara for residents and visitors. However, the Karrayu, like the other pastoralists in the Awash Valley, avoid fish.

Fentalle Mountain
Another physical feature of the Karrayu land is Fentalle Mountain with an altitude of 2007 metres above sea level, which makes it the highest elevation in the vicinity. The mountain is situated to the north of Metehara town. Fentalle Mountain, which has been formed by centuries of volcanic eruptions, dominates the scenery of the area. In addition, the name Fentalle is widely used as a personal name by the Karrayu. The main reason for this is to encourage the Karrayu to feel loyal to their country, starting from their childhood, so that they will care for the local resources and maintain them for the next generation.

3.2.3 Micro-climate and seasons

Since the district is located in a lowland area in the middle of the Rift Valley region, it is characterized by a hot climate (arid and semi-arid). The annual maximum temperature ranges from 32°C to 42°C while the minimum temperature ranges from 9.6°C to 22°C. The mean annual rainfall as shown by meteorological data for the area for the last twenty years is about 631mm per annum. However, the rainfall received in recent years is below the long-term average. According to Jacobs and Schloeder (1993: 12–14), this limited amount of annual rainfall is a result of being located within a low-pressure convergence zone. Though there is both temporal and spatial variability in rainfall, there are two rainy seasons. The main rainy season in Karrayu, as is the case in other parts of Ethiopia, falls between June and September, and is a result of very wet winds from the Atlantic and Indian Ocean meeting over the highlands. This season is locally referred as *ganna*. The short rainy season, referred to as *afraasa* is caused by monsoon winds from the Indian Ocean; it usually begins in March and continues to the end of April. The rainfall in this period is light and unreliable.

Figure 4: The short season rainfall pattern of Fentalle Woreda (1985–2009)

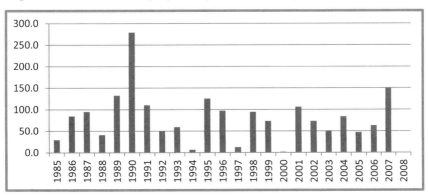

Source: Data compiled by Ethiopian Meteorological Agency (EMA)

The Karrayu have been rearing livestock for centuries and know how to deal with the spatial variation of pasture and water resources. They keep a mix of livestock species which have different grazing, watering and mineral needs, in order to maximize benefits and minimize risks and to adapt to the arid environment. This livestock mix consists of cattle (*horii*), sheep (*hoollaa*), goats (*rae'e*), and camels (*gaallaa*). The Karrayu have recently started to focus on camel herding as a coping mechanism in the light of diminishing pastureland and depleted water resources, since camels can survive for a relatively long period without water and they can travel long distances in search of water and pasture.

3.3 METHODOLOGICAL STANCES: QUALITATIVE APPROACH IN GEOGRAPHY

The methodological foundation of my research is two interconnected strands of thought: qualitative and participatory approaches as explicated by Creswell (Creswell, 2003; 2007). Participatory research as a branch of qualitative geographical research is informed by a critical social science paradigm (Bailey, 2007). Critically oriented qualitative inquiry appreciates that processes of knowledge production (including the debate on human-environment interactions) is political and situated in power relationships. In more specific terms, it is conscious of the mechanisms involved in approaching people and places that are outside the researcher's own community in order to avoid mistreatment and domination during the research processes (Clifford and Valentine, 2003; Dowling, 2005; Kindon, 2005; Willis, 2007; Best, 2009). Qualitative research methods in geography are broadly categorized into oral, textual and participatory methods. Conversely, some authors treat participatory research separately from qualitative research in terms of origins, philosophies and methods (Mayoux, 2006).

Despite similarities in terms of their philosophical and practical points of view, I treat the two separately for the sake of clarity. Based on these approaches, this work relies on a mix of methods and multiple sources of data, sometimes referred to as triangulation. The multiple sources of information make it possible to check for similarities (and variations) in the information provided.

3.3.1 Qualitative approach

A qualitative research approach has a number of advantages. It is useful for describing complex phenomena as they are situated and embedded in specific social, political, institutional and environmental contexts, and it has enabled me to study the dynamic interaction of local actors with their broader social and physical environment. I found that qualitative approaches were responsive to local situations, conditions, and stakeholders' needs, and to the changes that occurred during the period of the study, thereby allowing shifts in the original focus of the study to incorporate emerging new realities in the field (Johnson and Onwuegbuzie, 2004).

However, a qualitative approach runs the risk of being time consuming and there is a higher chance that the results will be influenced by the researcher's personal biases and idiosyncrasies. Irrespective of these factors, Creswell (1994) has underlined that there are several potential aspects of qualitative research which make it one of the favoured methodological approaches. In his view a qualitative approach: 1) necessitates fieldwork, allowing the researcher to observe behaviour and conditions in a natural setting; 2) considers the researcher as a primary instrument for data collection and analysis; 3) focuses on description because understanding is gained through words, pictures and other media; 4) is concerned with processes rather than products or outcomes; 5) is concerned with meaning, (i.e. how people make sense of their lives, experiences, and structures of their world); and 6) is inductive in nature, i.e. the researcher builds abstractions, concepts, hypotheses and theories from details observed.

3.3.2 Participatory approach

Within the framework of qualitative research, the study developed a set of participatory approaches. Conventional questionnaire methods suffer from problems such as identification of important research issues and their relevance to local people, invariably large numbers of questions, and the long time period required to administer the questionnaire. In contrast, short field visits are full of biases and may misguide researchers into believing they have seen an accurate picture of the field reality (Pretty and Vodouhe, 1997). These biases can be categorized into: spatial biases, time biases, people biases, and project biases (Chambers, 1983). Owing to these flaws in conventional approaches, there has been a recent rapid expansion of participatory approaches (Pretty and Vodouhe, 1997).

Regardless of the different terminologies used to denote participatory approaches, they are all interlinked through a set of common principles (Pretty, 1994): 1) defined methodology and systemic learning process, 2) multiple perspectives, i.e. the objective is to seek diversity, 3) group learning processes, i.e. recognition that the complexity of the world will only be revealed through group inquiry and interaction, 4) context specificity, 5) facilitating experts and stakeholders (i.e. role of the "expert" is best thought of as helping people in their situation to carry out their own study), 6) leading to sustained action, i.e. the learning process leads to debates about change which positively influence perceptions of the actors and their readiness to contemplate action. Implementation of the participatory approach in the field required a number of activities. The selection of specific study communities was done in collaboration with the local people. Introductory workshops were conducted to discuss the study design and its implementation process. The use of participatory methods offered the opportunity to implement a number of practical research tools, such as social and resource mapping, trend analysis, wealth ranking. The diversity of these tools helped situate the research in the past, present and future. As part of these participatory approaches, group and individual interviews, community and focus group discussions were organized to collect specialized information from specific gender and wealth groups who have different experiences with regard to the research topic. It was a complex process to get all these diverse groups together within the rigid social and cultural contexts that characterize the Karrayu pastoral community.

3.4 THE RESEARCH PROCESS: PRELIMINARY VISITS AND INITIAL ACQUAINTANCE

The research was conducted in three rounds over a period of eleven months from February 2009 till December 2011. Reconnaissance or an overview study through preliminary village visits was conducted during the first three months (February-April 2009). This gave me an opportunity to talk to a variety of community members across Fentalle *Woreda*, and these discussions brought my attention to a wide range of topics and issues. A deliberate attempt was made to elicit the views of various pastoralists in Fentalle *Woreda*. The reconnaissance visit covered a total of ten villages. Most village visits were informal and were made without any prior notice to the village. Together with my field research assistants, I walked into these villages and spent several hours talking to village leaders and other villagers, meeting the full village committee in some of the villages, visiting their homesteads and looking at the village records.

I realized that my observations during the first stage in the field required further verification at household level with different groups within the Karrayu community. Accordingly, the second round of the field research took longer and was conducted from March to September 2010. During this phase a variety of research tools were utilized to conduct the research. After selecting the study sites during the first cycle of the field research, during the second round I used various participatory tools such as resource mapping of the *Woreda* and wealth ranking, with the help of key informants, and identified four categories of pastoralists. Based on these categories, fifteen Karrayu pastoralists from each category were selected (i.e. sixty from each study site with a total of one hundred and twenty pastoralists) to further probe into the various topics. The participatory wealth ranking helped me to come up with differentiated groups of pastoralists and tailor my other research tools accordingly. It also gave me an idea of the meaning of wealth in the Karrayu pastoral community.

3.4.1 Selecting the units of analysis and study villages

Though I visited other pastoral areas of the country, such as Afar and Borana while I was working at the International Livestock Research Institute, my close acquaintance and previous rapport with the Karrayu community were good reasons for me to conduct my research among these pastoral groups. This research is a bottom-up study of pastoralists-environment-state interactions and how the Karrayu are using their agency in order to adapt flexibly. It also investigates new forms of resource access that help them manage risks and secure livelihoods under specific local contexts. When conducting research, it is necessary to fix limits both temporally and spatially within which to examine these complex processes of human-environment interactions.

In order to understand the risk management and livelihood practices of the Karrayu, I use two units of analysis: the homestead (locally termed *ganda*) and the individual pastoralist. Since homesteads are organized into communities, I also investigate how their members interact within, and with individuals outside, their homestead. I consider various livelihood activities, variations in mobility patterns, and employment of different strategies as processes negotiated among members, who use their knowledge of local ecology and resource accessibility, and exchange resources and assets through endowment, reciprocity or trade (Giddens, 1979; Sen, 1981; Long and Long, 1992; Arce and Long, 2000) to maintain or enhance livelihoods and manage risks arising from various sources. These risk management and livelihood practices pursued by Karrayu pastoralists are in turn influenced by changes in resource availability, access and the institutions that govern the management of the resources. Furthermore, Karrayu pastoralists depend firstly on the integration of individual pastoralists into a flexibly organized whole, and the homestead's articulation within the larger spheres of extended family and community within and beyond their physical environment. Even though I covered more than ten villages out of the eighteen *Kebeles* in Fentalle *Woreda*, for in-depth research and observation I concentrated on Giddarra and Illala villages to address some of the specific questions that I have raised. The selection of the two villages was based on the following criteria:

1. The type of livelihood activities that are practised in the two villages. In this regard, Giddarra is a village dominated by agro pastoral activities, whereas Illala is dominated by exclusively pastoral activities. This distinction is important for understanding variations in adaptation strategies between the localities and what contributes to these variations.

2. The physical location of the two study sites was also considered, in order to be able to study the interaction of Karrayu pastoralists with other neighbouring groups. In this regard Giddara village was chosen in order to see the interaction of Karrayu pastoral strategies with Ittu and Arsi Oromo farmers who reside in the nearby villages. On the other hand, Illala village was selected in order to see the interaction with neighbours such as the Argobas. This criterion is helpful in situating pastoral agency within its social context beyond the bio-physical environment.

3. The third criterion was the type of livestock species that dominate in the study sites. In this regard, camels are mostly raised in Illala village, while in Giddara village a mix of small ruminants is raised. This selection criterion was helpful for understanding the impact of the specific location of the villages on the livelihood strategies.

3.4.2 Specific research methods employed

In this sub-section, I outline the procedure of my research in the Karrayu pastoral community of Upper Awash Valley. I introduce the methods used for data collection and analysis, the informant and study area selection procedures followed, and the participants involved. The methods outlined in this section are in line with the philosophical orientation discussed above. At the beginning of my fieldwork in the Karrayu villages, I mainly used group-orientated participatory techniques to address the first specific research question. As the research progressed, I predominantly used interviewing and participant observation techniques to examine individual and general community issues and concerns and their relationship to climate-related problems. This approach enabled the participants to express their own constructions of climate-related vulnerability. It also allowed participants to emphasize the socially-orientated root causes of climate problems, making the research less prescriptive. The entire research process generally flowed in one of two directions: from discussion of 'non-climate' stresses towards linking these with climate-related problems, or vice versa. This was parti- cularly effective, given the particular climatic situation of the communities I visited, as it better facilitated a true representation of the perceived priority of climate stress. Rather than beginning with explicit questioning regarding climate stress and related problems which may paint a somewhat skewed picture of local concerns, participants were able to communicate in their own way the 'multiple stressors' influencing their vulnerable situation.

Participatory Mapping

In this study participatory mapping is utilized to generate information on the changes in resource availability. Through this tool it is possible to document the trends in the dynamics of water and pasture availability over the past several decades. For instance, during my field research I employed trend analysis and problem tree analysis to help my informants to pictorially represent the historical trajectories of development interventions that were behind the transformation of their resource base in the Metehara plain. This exercise of participatory mapping was useful in capturing the broader context in which the risk management and livelihood practices of the pastoralists have evolved over time. In addition to documenting changes in resource availability, participatory mapping proved very useful in documenting societal dynamics such as changes in customary institutions, migration patterns over the years, and how livelihood practices have evolved.

Intensive village case study

This study employs an intensive case study approach. A case study can provide a 'thick description' or analysis of a community's own issues, contexts and interpretations (Stake, 2005). This research is concerned with what can be learned from a particular case (Stake, 2005) – here, the Karrayu pastoral community's interactions with the various Ethiopian regimes and their environment. Because of the need to 'get close' to participants in order to understand their context-bound perspectives, I chose an intensive case study approach rather than a comparison of multiple case studies (Gerring, 2007). As part of the intensive village study, I have also the transect walk method in Fentalle *Woreda* to observe and collect place-specific data on farm lands, pastures, watering points and settlements. It is not my intention to generalize the findings from this particular case study, but rather to use the findings to address and contribute to larger questions and issues in climate change adaptation and human geography (Hardwick, 2009).

Individual/household level data collection

Under the qualitative umbrella mentioned above, I used a combination of semi-structured interviews, open interviews and opportunistic discussion as part of participant observation with community members. This combination of methods enabled flexibility – some methods were better suited to certain groups or individuals in the community than others. The combination of methods enabled a triangulation of findings, which is essential to the robustness of data and validity in qualitative research (Patton, 2002). Interviews were both semi-structured and unstructured. By interview I mean a more formalized context, where either the discussion had been organized beforehand, or the discussion was opportunistic but lengthy, in-depth and concentrated. Interviews were recorded by hand-written notes, depending on appropriateness. A field diary entry was written following each interview, including personal reflections on participant attitude, interview context, people present, questions asked and points to follow up. The advantage of a semi-structured format was that I was able to focus and direct discussion around specific content relating directly to the research objective. The disadvantage, however, was that my role tended to be 'interventionist' and there was therefore less room for new topics of inquiry to be revealed (Kitchin and Tate, 2000; Dunn, 2005). Unstructured interviews allowed more room for participants to express their personal perceptions and histories (Kitchin and Tate, 2000; Dunn, 2005). Personal accounts of significant climate events were a particularly effective way of drawing out the factors shaping vulnerability. Unstructured interviews were participant-led – questions asked were determined by participant responses. Importantly, unstructured interviews can allow perspectives to come to the fore that may be concealed by the dominant view (Dunn, 2005).

Interviews were carried out either with individuals or small groups of two to four people. Patton (2002) points out the advantages of unstructured group interviews in fieldwork (rather than 'focus groups'): often participants feel more comfortable when together than in an intensive one-on-one interview situation. Commonly, family and friends would come and go, contributing intermittently throughout the interview.

The intention was for the interviews to remain conversational, a format with which participants were most comfortable. The participants were able to emphasize the topics that they felt were important. The flexible nature of the interviews meant that I could explore topics about which participants were particularly knowledgeable. For example, some participants had specific knowledge of traditional weather forecasting techniques, while others were knowledgeable about food preservation. Often, a participant's specific area of knowledge acted as a starting point for the conversation, and discussion would branch out from there. Both semi-structured and open interviews allowed for unexpected topics and issues to come to the fore and be explored in more depth. With a few exceptions, I interviewed participants several times in order to follow up on points that were unclear or required more discussion.

A disadvantage of semi-structured and open interviews is that questions and responses are not standardized and directly comparable. This makes analysis more difficult as responses can be lengthy and convoluted (Patton, 2002; Overton and van Diermen, 2003). I participated in community life in as much as was possible within the bounds of my fieldwork term. Developing personal relationships with people is an important aspect of participant observation (Patton, 2002). During my fieldwork I lived in the homes of local families, participated in regular household routines and chores and took part in community activities. For example, I was given the assignment of taking the livestock to nearby watering points on a day-to-day basis. Due to the dryness of the area, daytime temperatures were very high during most of the time of my fieldwork. The majority of the households conducted their livelihood activities in the early morning (before 11) and late in the afternoon when the sun is about to set. Through this participation I formed relationships and increased my sensitivity towards participants' 'life worlds'.

Patton (2002) contends that in participant observation there is little distinction between 'interviewing' and 'observation' because the researcher is fully engaged in experiencing the situation. Informal, opportunistic discussions undertaken whilst participating in normal, everyday community life formed an integrally important part of my data, alongside more formalized interviews. These discussions enabled a closer contextual understanding of the way in which local people 'see things' than more formalized interviews. Data was recorded via field notes, when appropriate. Often, taking notes was not appropriate however, as this would have disrupted the 'normality' of my participation in a situation (see also Cook, 1997; Kearns, 2005). I kept a detailed and structured field diary where I recorded the contexts of participation, my recollection of discussions, observations, ideas and reflections upon my interpretation of situations (Cook, 1997; Kearns, 2005; Dowling, 2005).

Sampling

During my fieldwork, participants were selected using two types of purposeful sampling methods: *snowball* and *opportunistic* (Patton, 2002). For interviews, I aimed to involve participants who were particularly knowledgeable about the community as a whole and its history, for example elders, *abba gadas*, and Imams. I used snowball sampling to identify these participants. As 'changes over time' became an important point of discussion in my research, I invited a large proportion of older participants to contribute. However, I made a particular effort to also include younger participants as they often had a different perspective on aspects of socio-cultural change. A breakdown of participants by approximate age category can be seen in the table below. It should be noted that most participants did not know their exact year of birth. I used an opportunistic sampling method in conjunction with snowball sampling. Opportunistic sampling allowed the flexibility required for participant observation, enabling me to take "advantage of whatever unfolds as it unfolds" (Patton, 2002: 240).

3.5 MIXING VARIOUS INTERVIEW TECHNIQUES

Interviewing is widely known as the most common method of data collection in social science research. Types of interview may range from informal and unstructured to semi-structured and structured. While we often tend to use one type of interview as the dominant method in our research, we often use other types throughout the research period either consciously or unconsciously. Bernard (1988) has discussed four types of interview techniques which he termed as interview control characterized on the basis of the interview situation and the amount of control exercised on the responses of the informant. They are: informal interview (absence of structure or control), unstructured interview (clear plan and minimal control), semi-structured interview (use of interview guide) and structured interview (response to a fixed set of questions). However, irrespective of their type, all interviews involve interactions, thereby subjecting the different processes of interviewing to a similar set of dynamics. Another important aspect is the appropriate use of these interviews, which is dependent on the duration of the research and the specific context within which the research is conducted.

I used purposive sampling in the study area to begin interviewing people who engaged in particular types of livelihood activity. In small villages, this is not difficult. Social scientists use purposive sampling when they have identified people who are likely to offer valuable information about the research topic (Bernard, 1995:95). Following the identification of interviewees through purposive sampling, I used snowball sampling to acquire other interviewees. According to Bernard, a snowball sample is useful when used in a relatively small population, working with people who are likely to know one another, a logical fit in the context of the Karrayu. In a snowball sample, the researcher identifies one or more individuals and requests the names of other likely participants for the study (Bernard, 1995:97). It was not unusual for me to go to someone's home to conduct an interview, and to hear from him or her about another person who might be good to talk to.

The three types of interview were used to varying degrees during the research. Mainly, I used a two-pronged approach to the use of interview methods. First, I used different types of interview at different stages of my research, for instance informal interviews in the initial phase of the research, and unstructured interviews in the subsequent phase. Most of the informal and unstructured interviews were conducted during the reconnaissance survey and other preliminary interactions with individuals. Second, I used semi-structured and interviews after the study villages were finalized and I had built a good rapport with the Karrayu pastoralists and other people in the area. Since semi-structured interviews and focus group discussions were more extensively used, I will discuss both these techniques briefly.

Key informant interviews

Key informant interviews were conducted with ten community elders between March and August 2009. The purpose of these interviews was to provide general contextual information, direct me towards particular data sources, gather viewpoints and better understand community-based data (Patton, 2002). Key informant interviews provide the background to this research by identifying the major historical events in the Karrayu pastoral community over the past seventy years. For this purpose I relied on community elders who have a depth of knowledge regarding events of significant value to the Karrayu pastoral people in Upper Awash Valley. I conducted the key informant interviews as part and parcel of the participatory tools that are used to generate historical data. The data generated using this particular technique is presented in Chapter Four in an attempt to locate the research problem in a broader context and thereby answer research question number one.

Photo 1: Key informant interviews with Karrayu elders

Source: Author

Semi-structured interviews

Semi-structured interviews are a central part of all participatory approaches in which an interview guide is employed (Bernard, 1988). The questions asked are content-focused and deal with areas judged by the researcher to be relevant to the research question (Dunn, 2005). In this type of interview the role of the researcher is recognized as being facilitative, and he or she may redirect the conversation if it has moved too far from the research topic (Dunn, 2005). I used semi-structured interviews to collect data from the agro-pastoralist community members based on a set of pre-determined questions. This interview method had a number of advantages over other methods, especially within a time-bound research project like this one in Fentalle *Woreda*. Since an interview guide was used, the risk of accumulating a high volume of gap information could be minimized. The entire interview was conducted in a discussion mode which allowed for refocusing the interview on some of the emerging issues. In other words, semi-structured interviews were not necessarily limited to the interview guide. Moreover, the interview guide contained several lead questions that were critical in initiating a dialogue with the participants. In designing and conducting the semi-structured interviews, I was aware of the fact that if not properly designed and conducted in the field, semi-structured interviews that are restricted to the questions in the interview guide may leave many gaps in the information gathered. Moreover, I was always careful not to overemphasize the interview guide, which would have resulted in the participants losing interest in the entire process.

Focus group discussions

Focus group discussion is a technique whereby a group of people (as few as 6 and as many as 30) is brought together for a joint interview session (Bernard, 1988). Bernard (1988) explains that focus group discussion is not a way to measure precisely the amount of some behaviour in a population. But it is an excellent method for getting an indication of how pervasive an idea, value or behaviour is likely to be in a population, or for understanding how deeply feelings run about issues. In other words, focus group discussion may be defined as an interview style designed for small groups where the researcher strives to learn through discussion about conscious, semiconscious, and unconscious psychological and socio-cultural characteristics and processes among various groups (Berg, 2004). Thus, focus groups allow the researcher flexibility, scope for observation of interactions, collection of substantive content within a limited time frame, and access to various sub-groups within the community (Berg, 2004).

Six focus groups at the two study sites were conducted at different stages of this research in order to gain critical inputs from the community. Some focus groups were used to direct the research to the community level and also to present and verify the research findings generated using other methods. In the Karrayu community, which is characterized by differences based on wealth and gender, I found focus group discussions to be a very effective method. In such circumstances, it was not politically pragmatic to talk to the community members as a whole, because I was not certain about who would feel comfortable in whose presence. Also, in large group settings, the powerless in the community often remain silent or make censored statements. Women and resource-dependent poor often constitute this group. Therefore, focus group discussions were used to gather information from different sections and sub-groups in the study communities; these primarily consisted of women, and different wealth groups in the community.

Even though focus group discussions are an effective method of data collection, they are not free from problems. Talking to several smaller groups in a hierarchical community always runs the risk of leading to controversies and confusions. Dominant groups may doubt the intentions of the researcher behind what they may see as "secret talks" with certain groups. Moreover, interacting with women in a typical male-dominated pastoral community can be challenging. However, my experience in this regard was very positive, as I was able to conduct several meetings and discussions with groups of women. When special meetings were organized, women came in good numbers and articulated their views clearly. In this regard, my long-term involvement with the community was an asset.

3.6 FIELD NOTES AND PARTICIPANT OBSERVATION

Fieldwork or participant observation is the key data collection method (Hammersley and Atkinson, 1983). There are different kinds of participant observation along a continuum (Hammersley and Atkinson, 1983): *complete observer*: where the researcher does not interact with people in the setting or participate in any of the activities, but is there strictly to observe; *observer as participant*: where the researcher is there primarily to observe, but will participate in activities or interact with people only when necessary, and in a minimal capacity; *participant as observer*: where the researcher seeks out participation in the setting specifically for the purposes of the study, interacts with others and participates fully in activities, and uses that participation as observation; and *complete participant*: where the researcher is an existing, active member of the setting, interacts with others and participates fully in activities, and also uses that participation as observational data. The type of fieldwork used in this research was the *participant as observer* type, with interactions with households and participation in some activities. During the major part of the field research, I was able to spend much of my time among the Karrayu pastoralists in the study villages, and I was given the task of watering the small livestock at the nearby pond. This helped me to develop a rapport and closely observe the everyday activities of the individuals as they interacted with their environment and with each other. It also helped me to conduct interviews in a very relaxed atmosphere in the villages. This participant observation was supplemented by interviews.

While in the field, I tended to not take notes during informal events because I thought it would make people behave oddly. I could not imagine carrying my notebook to a farm plot where people were watering their field, or taking notes while herders were taking their animals to a watering point. I did take notes where it would not be unusual for others to be writing as well. My main concern was for people to feel comfortable around me and I believed that carrying a notebook around all the time was not the way to achieve that. During my fieldwork activities, I noted key pieces of information on the spot and returned to the house as soon as possible after the activity to write up a more complete account. Using the *contemporaneous field note taking* approach described in Emerson, Fretz and

Shaw (1995: 13), I wrote highly descriptive accounts of what I had seen and heard. These field notes concentrated principally on the events taking place, the topics and ideas being discussed, the decisions being made, by whom, and the logic and/or criteria used for making those decisions. I also documented the thoughts people shared (using word-for-word quotes when useful and possible), their actions, responses, and interactions in relation to these events and decisions, taking into account body language when appropriate.

3.7 DOCUMENT ANALYSIS AND SECONDARY INFORMATION

A detailed analysis of relevant state policies and laws concerning the management of resources in the valley was undertaken. This policy analysis tried to capture the overall policy environment within which conservation and management of dryland commons take place. It provided an understanding of the processes and strategies through which community-level resource management arrangements in Fentalle were shaped and influenced by external factors, including state laws and regulations. It brought out various historical and current trends in the management of natural resources and livelihoods in the valley.

3.8 ANALYSIS AND WRITE UP

Throughout this dissertation I followed a very flexible strategy of data gathering, write up, and data gap filling. In this regard, organizing my field research in three rounds was helpful. So, the overall data collection and write up process followed an iterative and flexible process where data is collected and analysed in the field and issues that emerge feed back into the entire research process (Hickey and Kothari, 2009). Data analysis was an on-going process throughout the fieldwork, assisted by the participants themselves. Emerging themes were discussed and evaluated with the participants. From this, further important themes could be identified and investigated. Accordingly, the last round of my field research was utilized to both fill in gaps and confirm the data generated during the major second round of field research.

Most interviews were summarized and translated from Oromiffa to Amharic. Throughout the field research I relied on my assistants who themselves are Karrayu for an explanation of specific local terms and the translation and summary of the interview materials. The hand-written interview notes, field notes from opportunistic discussions and field diary entries from participant observation were analysed manually. Analytical insights and interpretations that emerged during data collection formed the organizational basis for distilling the data into key themes (Patton, 2002).

3.9 ETHICAL CONSIDERATIONS

The consent of research informants is one of the important ethical considerations that one has to bear in mind during the research process. This consent involves agreements on the part of the informants on several fronts such as the allocation of their time for the research and possible consequences that may arise due to their involvement in the research. In other words, research participants should be given as much information as might be needed to make decision about whether or not they wish to participate in a study. To begin the field work and collect data, it was necessary to get permission from the government officials. Hence, I started seeking permission at *Woreda* level administration. After briefly explaining the purpose of my research (that it is part of my research program and it is for the completion of my PhD degree) and objectives, and possible questions that I can ask both the government officials and people from local community, permission was granted. I explained for the Fentalle *Woreda* administrator the government sectors I need to talk to, those who have a direct link and relevance to my research. Following this, the *Woreda* administrator wrote me a letter of permission and assistance for all the eighteen *Kebele* administrators in the *Woreda*. It was after getting this permission from the *Woreda* administration that I started seeking the informed consent of my informants. All of my informants and focus group discussion participants were informed about the overall purpose of the research, and that their participation was voluntary. Accordingly, all my informants were not obliged to respond to questions that they feel not comfortable to reply.

Another important ethical consideration was keeping the names of my informants anonymous. During any field interaction the participants identified themselves by their individual names. Self-identification by the participants was primarily seen as part of the local culture where it is customary to introduce oneself to outsiders by name. However, no individual participant was forced or motivated, in any way, to disclose her/his name if she/he chose to stay anonymous. As a principle the original names of the participants were protected by a measure of anonymity. The names of the participants were only used in order to clarify or verify data during the field research period. In the thesis and other published documents a two-pronged approach was adopted to deal with the identity of the participants: (1) as a general ethical principle, pseudonyms were used for all women informants in conformity with local cultural practices and as a measure of extra care given the sensitivities, (2) with regard to male informants, a mix of real names and pseudonyms have been used under the condition that verbal consent was received in all cases where real names have been used.

Another ethical challenge during the field research was the locals' perception of the researcher as rich and powerful. People were unhappy when told that there would be no immediate benefit from my study, for they expected financial or some other kind of help. More often than not, I was asked why such studies are carried out if the findings are not used to provide solutions to the plight of pastoralists. It was very difficult to answer this question and it is an outstanding ethical problem as far as research in development studies is concerned.

The researcher is in a cleft stick when it comes to deciding whether he/she should remain detached from the situation (so as not to influence or 'bias' the course of events under study) or actively do something to relieve people's misery. Many argue that research has to be 'objective', and that engaging with the social problem under scrutiny would mean surrendering to the object of study, suggesting that the researcher should run away from this 'captivity'. In my opinion, research in development studies should not only be done for research's sake, and a strong link between research and practice in development studies is essential. During my stay in the field, people came up with puzzling questions like "If you have nothing in your hand why are you here?" or "We don't want your paper if it doesn't bring us food, cash or a water pump." These were some of the challenges that I faced. In some cases they thought of me like a government official and wanted me to pass their opinions to the bodies concerned.

However, despite all these challenges I managed to obtain the data that I needed, since I was born and brought up in the area and that allowed me to be flexible in accordance with the situation. In addition, the help of the Karrayu field assistants, with whom I had attended classes when I was in high school, facilitated the entire data collection process. These circumstances helped me to create a rapport with the people and obtain the relevant data. I tried to justify the relevance of this study to my respondents by arguing that its findings could be used later by policy makers in order to develop need-oriented and context-informed development projects.

4. LIVELIHOOD INSECURITY IN CONTEXT: HISTORICAL TRAJECTORIES

4.1 INTRODUCTION

In this chapter I explore the broad historical contexts that have shaped the risk management and livelihood security stategies of pastoralists in the lowland parts of upper Awash valley since the middle of the last century. Socio-political factors and processes underlie the vulnerability to climate stress of Karrayu pastoralists in a semi-arid environment, limiting the availability and smooth functioning of local adaptation strategies. This chapter outlines these historical processes, which are priority concerns in the community despite frequent climate-related stress. The structural factors and processes shaping vulnerability are largely a product of rapid social change, driven by distant processes. By social change, I mean the broad range of human factors that affect the pastoralists' livelihood security and risk management strategies, as opposed to the immediate biophysical factors influencing exposure. Social change thus encompasses social, cultural, economic and political factors and processes. Any endeavour to better understand the dynamics of risk management and livelihood security among pastoralists must start by exploring the dynamic and macro socio-political and ecological trajectories of change. I argue that these dynamics have critical implications for understanding contemporary approaches to risk management and livelihood security among the pastoral community.

Accordingly, the macro sources of vulnerability that have influenced, in one way or another, the livelihood security and risk management strategies of local actors in the semi-arid environment of Upper Awash Valley are explored. The data presented and analysed in this chapter was generated using various sources. Secondary sources of information such as government documents concerning the major activities in the Upper Awash Valley were consulted. In addition, primary data were obtained through focus group discussions and key informant interviews with elders, while various PRA tools such as problem identification and historical time line were utilized in focus group discussions. Furthermore, biographical interviews of selected informants were used in order to better elicit the historical trajectories of vulnerability, so as to better understand how the Karrayu pastoralists come to be where they are now. This chapter sets the foundation for tackling the three specific research questions that will be discussed in the subsequent three chapters.

4.2 STATE-PASTORALISTS RELATIONS: 'DEVELOPMENT' AND THE PERILS OF PLANNING

It is important to locate the pastoralists' livelihood insecurity within the historical context of state interventions that excluded the locals from processes of development in respect of their natural resources and disrupted the age-old pastoralists-environment relations. In line with the political ecology approach, this chapter shows that the risk management and livelihood practices of the Karrayu pastoral group are located within broader contexts of social, economic and political structures and institutions (Robbins, 2012). Such perspective helps us to frame local level processes and everyday livelihood struggles into broader historical contexts. These historical processes not only disrupted the pastoralists' relation- ship with their natural resources but were also instrumental in the production of environmental risks that have affected the livelihood security of various pastoral and agro-pastoral groups inhabiting the valley. Even though there are variations in their political ideologies and policies, all three regimes in Ethiopia in recent times share common characteristics when it comes to their vision regarding pastoralists in the country. While the imperial and socialist regimes had pictured pastoralist as a threat to national security and the resources located in these lowland areas as 'unproductive', the Ethiopian people revoluntionary democratic front (EPRDF) regime considers pastoral areas as fertile frontiers of development where investment in agriculture, dam construction and mining can boost the transformation of the country's economy. These processes of national development have led to the production of environmental risks that render the pastoralists' livelihood insecure. Pastoral communities have become the victims of state development interventions. As has been discussed in the theoretical section of this book, the political contexts that have shaped the utilization and management of resources in the peripheral areas of the country are the major driving forces that have shaped ecological risks that influence the livelihoods of the pastoralists. In the following sections, I present the major development interventions in the Metehara plain and how they have affected pastoralists-environment interaction through the production of risks.

4.2.1 Commercial farms and loss of access to resources

Since the begining of the 20th century, the Ethiopian government has agressively incorporated the lowland peripherial areas of the country by force. This goes beyound territorial or geographical incorporation. The lowland areas of the country, which are mainly inhabited by pastoral and agro-pastoral groups, have been seen as unproductive areas that can be developed and used as a vehicle for realizing the government's idea of a modern economy. This could be clearly observed during the Haile Selassie regime, which followed capitalist ideas in modernizing the agricultual sector of the country's economy. During this period, large tracts of land located in the lowland areas of the Awash valley were given to private investors, and state farms were introduced which were further intensified during the socialist regime (Derg). These imported ideas of development, however, served the interests of the ruling aristocrats and nobilities. During the 1960s, the government of Ethiopia established the first large-scale commercial farms that led to the eviction of pastoralists from their priceless seasonal land and water resources. These were essential for pursuing their livelihood based on extensive livestock production. This modernist approach to development, which values a sedentary lifestyle over a pastoral way of life, and large-scale commerical farms over subsistence production, has degraded the local livelihood conditions of the people who have inhabited the valley for generations, and underestimates their idea of development.

Prior to the 1950s, the name Metehara might have meant little to many people in Ethiopia. But for Hendels-Vereeniging Amsterdam (HVA)-Ethiopia, the parent company of HVA-Metehara, the name acquired some significance. This part of the Awash Valley is one of the richest in terms of alluvial soil and its topography is suitable for modern crop and livestock production (Ayalew, 2001). Its abundant potential for irrigated agriculture started to be recognized and appreciated after the end of 1950s and beginning of the 1960s. The construction of Koka dam in 1962, and the introduction of the first commercial farm in the Upper Awash valley in 1952, signalized the interest of the imperial regime in the land and water resources of the valley. In the subsequent decades, the valley, which was predominantly inhabited by pastoralists and endowed with potential for various interventions such as irrigated agriculture and hydro-power generation, caught the eye of state and private enterprises. During the 1940s pastoral land had been reclassified as 'public domain', which permitted the state to give 'unused land' (land not used for agriculture) to individuals and groups.

In order to put its ideas into practice, the imperial regime established the Awash Valley Authority (AVA) in 1962 with the mandate of planning development and adminstering resources in the Awash valley (Ayalew, 2001). The primary vision of AVA was to optimally use the resources in the valley through the establishment of large-scale commercial farms that would grow crops such as sugar cane and fruits for industry, thereby modernizing the agricutual sector.

Figure 5: Cultivated area by crop variety in Upper Awash Valley (1970s)

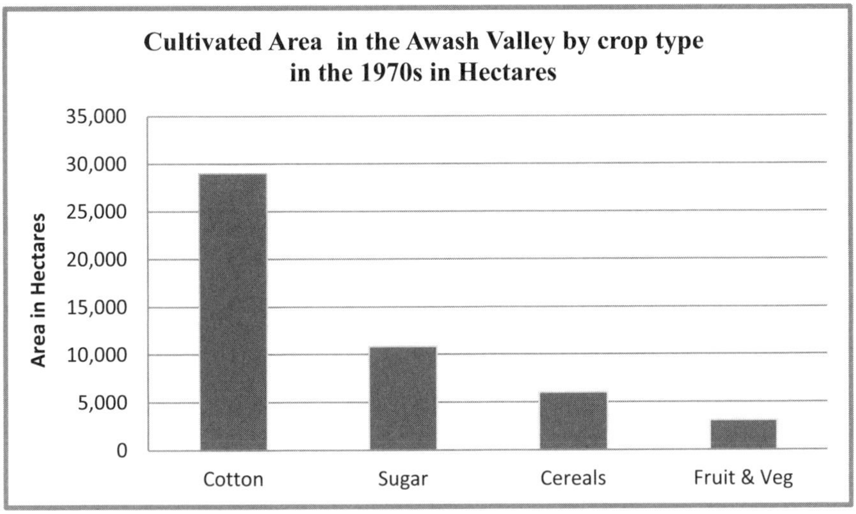

Cultivated Area in the Awash Valley by crop type in the 1970s in Hectares

Crop	Area in Hectares
Cotton	~29,000
Sugar	~11,000
Cereals	~6,000
Fruit & Veg	~3,000

Source: Based on Bondestam (1974)

The two sugar factories that preceded the MSF in the Awash valley were Wonji and Wonji-Shoa factories that were established in the Wonji plain in 1954 and 1960 respectively. In 1964 the idea of expanding the suagr industy to the Metehara plain became topical with a view to the increasing sugar consumption in Ethiopia and due to the fact that no further expansion of the Wonji and Shoa Sugar Estate in the Wonji plain could be envisaged.

Consequently, in July 1965 an agreement was signed between the imperial government of Ethiopia and the well-known Netherlands company Hendels-Vereeniging Amsterdam (HVA), under which HVA aquired a concession of 11,000 hectares of land for cultivation along the Awash River in the Metehara plain, taken from the prime dry season grazing area of the Karrayu pastoral groups. Complementary to this agreement, HVA-Ethiopa undertook to establish a new Ethiopian company, named HVA-Metehara, to which all rights and obligations under the Metehara agreement were transferred. Upon the establishment of HVA Metehara, a direct start was made with an extensive survey and thorough soil investigation, while experimental sugar cane was planted (Ayalew, 2001). These trials fully justified the high capital investement required to open the way for large-scale development of the area. Consequently, the MSF, with a crushing capacity of 17,000 quintals of sugar cane, became officially operational in November 6 1969. The factory started to produce 1,700 quintals of sugar per day. Here, approaching the research problem under investigation from political ecology perspective adds value by situating contemporary problems of risk management and livelihood security in their complex global-local interlinkages. During the middle of the last century, Ethiopia had attracted large number of foeign companies that were engaged in investment in agricultural sector. Many of these investments, that were mainly export-oriented, had happened in the lowland areas of the Awash valley. The foreign investors that had involved in joint-ventures were from British, Dutch, Israel and Italian.

Photo 2: Partial view of the Metehara Sugar Factory and plantation

Source: Author

These large-scale investments in the agricultural sector were based on the assumption that the fertile grounds of the Awash valley were unproductive and needed to be transformed in order to benefit the country. The interests of the Karrayu pastoral groups who relied on the resources in the area for their livestock-based risk management and livelihood practices were ignored (Hogg, 1997; Ayalew, 2001).

4.2.2 Conservation without people

Another intervention by the Ethiopian state that materialized in the name of 'safeguarding nature' in the middle of the 20th century was the establishment of a national park. The establishment of conservation areas in the lowland areas of the country were influenced by global capitalist discourses of nature that valued economic interests, scientific research and recreational values over indigenous people's customary rights and knowledge of resource management. The establishment of Awash National Park in the Metehara plains in 1966 also involved evicting the Karrayuu pastoralists and keeping them away from their natural resource base in an attempt to preserve the natural beauty of the landscape. This was enforced by a fines and fences policy on the part of the Ethiopian regime. No consideration was given to the Karrayuu pastoralist communities who used the resources as dry and wet season grazing sites that were crucial to their risk management and livelihood practices.

The establishment of Awash National Park (ANP) in the Metehara plains was one manifestation of the nature of state-pastoralists relations in the middle of the 20th century. The principal logic behind such conservation practices was based on a separation of nature and society, and not paying due consideration to the participation of the locals in the planning process and management, in order to let them benefit from the national park. There was complete negligence of the locals' knowledge of ecology and their relations with the fauna and flora located in the conserved area. The area where ANP was established was substantially grazing land in the possession of the Karrayu. The Karrayu were neither consulted nor compensated, then or since, and all that they were told was to move out of the area. No type of benefit-sharing arrangement or community participation mechanisms have ever been instituted to allay their fears and to ameliorate the losses sustained by the communities which suffered from the loss of such prized grazing land as the Awash National Park. This has created a situation where neither the inhabitants of the park, nor the wildlife, nor the communities with their livestock are beneficiaries of the park, for different reasons (Ayalew, 2001).

All in all, the conservation of natural resources did not take into consideration the locals' relation with the natural resources, and priority was given to nature conservation over the people. It is still in the memories of the Karrayu elders that one year before the establishment of the Awash National Park, government troops came to the area and burned down adjacent villages, which are located in the foothills of Mount Fentalle. The Karrayu pastoralists who enter the newly fenced park in search of pasture for their animals are forced by the park authorities to pay heavy fines, depending on the type of animal that is found in the park territory. In some cases, the scouts of the park have fired on and killed the livestock when they attempted to escape. Furthermore, the establishment of the park has resulted in the loss of key watering points used by the Karrayu for their camels, and denial of their dry season retreat and migration areas.

Figure 6: Dry-season pasture land lost to various interventions (in hectares)

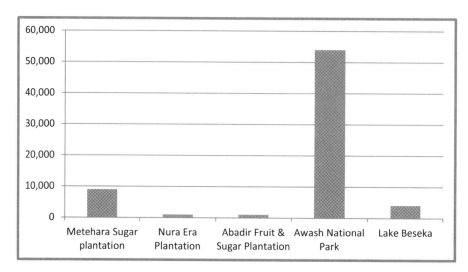

Source: Chart based on Bondestam (1974), Ayalew (2001) and MSF

Similar to the establishment of fortress conservation in many other African countries, the establishment of the Awash national park did not take into account the co-existence of their livestock with the wild animals, and that it was established on the basis of the 'Western' assumption of a nature-culture dichotomy (Neumann, 1998; Bryant and Bailey, 1997).

While the Ethiopian state claims control over the territory for conservation and economic purposes, the Karrayu pastoralists consider the acts of the state as a threat to their livelihood and survival. The Karrayu object to the state's approaches to conservation, building their argument upon their long-term traditional knowledge of nature conservation that is part and parcel of their worldview. For instance, a Karrayu elder explains that, "We always saw the oryx, gazelles and deer with our good eyes – we never tried to treat them badly. You know…they used to mix up and freely move with our livestock. It was taboo to harm these wild animals that have no one to take care of them. If we did something, we knew that Waqqaa would punish us by killing our livestock" (Abba Shantema: Metehara, June 2010).

Assigning the Karrayu range land as a conservation area was not accepted by the locals and there have been serious conflicts between the park authorities and the Karrayu pastoralists. Some Karrayu households have remained deep inside the park for so many years that they no longer believe they are trespassing within the boundaries of the park. In times of stress due to drought, the Karrayu are forced to enter the park territory. For instance, some households mention confrontations with the park authorities during the severe drought years of 1973 and 1974. The drought had led to a decline of their livestock and later to deterioration of the grazing resources outside the national park. The severe situation led the Karrayu to search for forage on the grounds of the park and thus to trespass within its boundaries. Following this the park authorities reported the cases to the then military regime and the pastoralists were evicted from the park. During the 1984 and 1985 drought, the Karrayu pastoralists reinvaded the park. This time they were allowed to occupy the slopes of Fentalle Mountain and the Ajotere area. Following the change of government in 1991, some pastoral households tried to take advantage of the political turmoil and the volatile situation by occupying land deep inside the territory of the park.

In the eyes of the Karrayu pastoralists, the establishment of both the sugar plantation and the Awash National Park have seriously affected their traditional strategies, which were based on seasonal migration of livestock between three grazing sites to reduce climate-related vulnerability such as drought due to the erratic nature of rainfall in the area. However, the establishment of the park is considered to have no benefit at all, compared to the sugar plantation, which at least provides them with grass and cane tops in times of severe drought. The sugar plantation also provides some support by hiring members of the local communities as guards. In the case of the park, the locals see no concrete value as a protector of wildlife, as their numbers keep on dwindling or as a provider of support for the Karrayu community.

Figure 7: The progression of entitlement failure of pastoralists

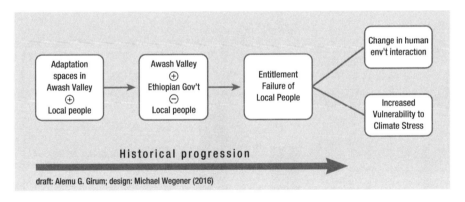

The coming of the state into the valley in the form of large-scale commercial farms is one of the major non-climatic human forces that have resulted in distortion of risk management and livelihood practices and hence increased the vulnerability of the local pastoralists. This is specially so through denial of access to key grazing and watering points for the extensive livestock rearing system on which the Karrayu pastoralists used to rely. The development policies of the then imperial regime, which was geared towards export in order to generate foreign exchange, seriously affected the 'pastoral spaces of adaptation' used by the Karrayu according to seasonal variations of rainfall and pasture.

What makes this state of affairs all the more harsh and outrageous is the fact that the land lost to the pastoralists was invariably the low-lying flatlands that the Awash inundated easily, making them highly productive and offering the best pasture for the livestock. These particular places were highly crucial for the adaptation strategies of the local Karrayu pastoralists. The damage inflicted on these communities tended to be irreversible. They were not nomads, but lived most of the year in close proximity to the river, in particular during the dry season which lasts some nine months of the year, from September to May. The government of that time had the power to decide on the configuration of access to resources and resource use practices by introducing large-scale commercial farms. These differential power relations determine who can have access to what kind of resources. They tilted the balance of power in favour of the government side, coercing the local Karrayu pastoralists to abandon their grazing land for reasons defined by the powerful actors. In line with the theoretical framework of political ecology which gives emphasis to power relations that determine human-environment interactions, the pastoralists have to face drastic ecological changes in their respective environments, suffering from severe overgrazing, erosion, circularly dwindling herds and subsequent pestilence and famine. In contrast, the national governments of the three regimes have continued to accumulate a huge amount of profit (see figure below) from the sugar cane plantation in which the local pastoralists have no share at all.

Figure 8: Annual growth of profit (in Birr) in the Metehara Sugar Factory

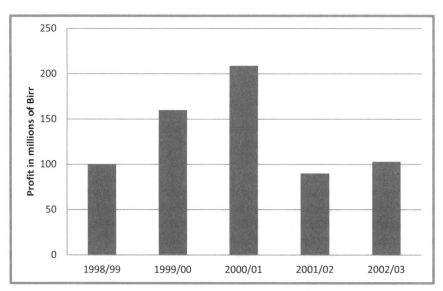

Source: Compiled based on MSF, 2005

In the context of climate change, O'Brien and Leichenko (2000) argue that globalization processes have inherently unequal implications for the well-being of different regions, countries and social groups. The process of economic globalization fundamentally affects who 'wins' and who 'loses' from the effects of climate change. They argue that those most marginalized by globalization are those likely to be worst affected by climatic changes due to restricted access to resources and assets necessary to reduce exposure to, and cope with, climatic changes and extremes (Olmos, 2001).

4.3 ENVIRONMENTAL SOURCES OF LIVELIHOOD RISK

Environmental risks are embedded in resources and relations that guide access to, and utilization of, these critical resources. As I have explained in the theoretical section of this book, the bio-physical environment in which the Karrayu pastoralists operate influences their forms of organization and resource utilization, as well as the adaptation activities in which households engage. In other words, the environment in a non-equilibrium dryland situation 'acts upon human wills and behaviours' (Latour, 2004). Since the effects among variables within an ecosystem are complex and intermittent, non-equilibrium dynamics are often the norm. This is particularly true for dryland ecosystems, where resource management is often highly complex, since it is not a matter of following up a single factor, but of grabbing opportunities and making flexible responses. For instance, Karrayu pastoral production systems are often interpreted as a direct coping response to environmental uncertainty, because intrinsic mobility allows people to respond flexibly to droughts and general resource scarcity in dryland ecosystems (Scoones, 1995). Though the natural environment has agency in influencing the will and behaviour of pastoralists in dryland ecosystems, here I argue for what Zimmerer and Basset (2003) have called a critical realist perspective of natural events (hazards).

4.3.1 Water scarcity, frequent drought and food insecurity

Research concerning the impact of climatic stress on livelihoods mostly adopts the 'realist perspective' by regarding hazards such as drought as something that happens to society. Such a perspective sees hazards and risks as objective, empirical realities with definable and measurable qualities. Hazards and the risks they pose have inherent qualities and effects. For example, droughts are treated not as Socio-ecological phenomena but as natural hazards that pose inherent risks and can be examined only in empirical ways. But as Wisner *et al.* (2004: 4) point out, such an approach "risk(s) separating 'natural' disasters from the social frameworks that influence how hazards affect people, thereby putting too much emphasis on the natural hazards themselves, and not nearly enough on the surrounding social environment". In contrast to this view, as this research shows, increased vulnerability to drought of Karrayu pastoral households is not because drought is 'happening' to them. Rather, it is the combination of several factors that are rooted in the socio-political structure. These have been compounded over the years and have disrupted the social fabric of pastoralism, thus rendering the pastoralists vulnerable to drought.

The Karrayu communities have been hit by severe drought several times in recent decades. The first period of drought was from 1972–1974; the second from 1980–1981 and the third from 1984–1986. According to my elderly informants, the area has also been visited by severe drought as recently as 1990 and 2002. It is estimated that in the 1984 to 1986 drought the Karrayu and Ittu lost 20% of their cattle, 7% of their camels, and 25% of their sheep and goats (Tibebe, 1997). A much bigger number of livestock and human lives are believed to have been claimed by the 1980 to 1981 drought (the one locally called the year of unprecedented cattle loss). It is still remembered as the most devastating, even if it is difficult to substantiate this with precise figures. With the help of the elders, it was possible to reconstruct a time line that shows some of the major droughts starting from the imperial regime. According to my key informants, drought is a frequent phenomenon that affected the area in the years 1958, 1964, 1969, 1974, and, during the Derg regime, 1979 and 1983–85, as well as recent droughts in 1994/95, 2002, 2008, and 2010.

Figure 9: The annual rainfall pattern of Upper Awash valley (1966–2004)

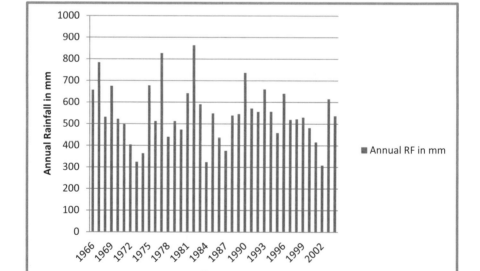

Source: Compiled from data from Metehara sugar factory, 2005

As illustrated in figure 9 above, the Upper Awash Valley area, where the Karrayu pastoralists reside, is by its very nature moisture deficient. The bimodal rainfall in this area has a very erratic nature and precipitation is very low. The timeliness of the rains is a crucial factor that determines the germination of grasses and the availability of water for both livestock and human use.

These recurrent and severe droughts have rendered Fentalle *Woreda* one of the most food insecure *Woredas* in the region. According to the *Woredas* Disaster Prevention and Preparedness Office (WDPPO), thirteen *Kebeles* out of the eighteen in the *Woreda* are categorized as food insecure and have required continuous support during the last three years (i.e. between 2007 and 2010) through the Productive Safety Net Programme (PSNP). The Productive Safety Net Programme is meant to build households assets and reduce destitution. As part of an early warning and drought management strategy, PSNP supports the beneficiaries either through direct support or through food-for-work activities, meaning they are involved in various public activities such as water harvesting and rural road maintenance. Annually, PSNP provides six months of support to households in food insecure *Woredas* of the country with the objectives of reducing household vulnerability, the improvements of household and community resilience to shocks and breaking the cycles of dependence on food aid, by "providing transfers to the food-insecure population in chronically food insecure *Woredas* in a way that prevents asset depletion at the household level and creates assets at the community level" (GoE, 2006). Accordingly, the majority of the beneficiaries in this program participate in labour-intensive public works activities such as soil and water conservation, water harvesting, small-scale irrigation, reforestation, rural infrastructure development, and water supply schemes. On the other hand, those households who do not have the ability to participate in the labour intensive activities of the program receive unconditional direct support in the form of cash and food. Fenatalle *Woreda* is one of the beneficieries of the national PSNP program. During the field work and through closer examination of the *Woreda* DPPO documents, it has been understood that the PSNP contributed a lot in consumption and protecting community and household assets. Protection of household and community assets reduces vulnerability to the impacts of climate variability and change. However, during the field work many of the informants mentioned that the selection of beneficiaries who are involved in the program created the problem of exclusion and inclusion within the Karrayu community while some of the households who are participating in the program develop a sense of dependency. Furthermore, the pre-condition of the public-work program that was designed based on the highland farmers' culture and work habit restricted the pastoralists from actively participating in it. The following table shows the number of beneficiaries involved in public work participation and receiving direct support from the Safety Net Programme from 2007 to 2009.

Figure 10: Number of PSNP beneficiaries in Fentalle Woreda (2007–2009)

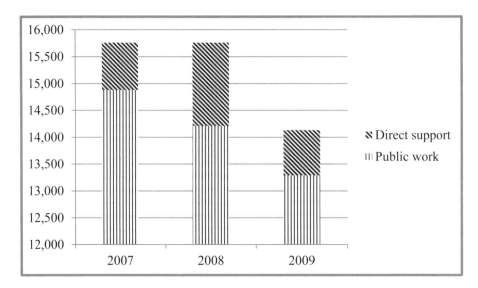

Source: Based on WDPPO (2010)

However, due to the recurrent drought that affects the area, relief aid is provided periodically through emergency aid programmes. The provision of emergency aid is necessary in order to avoid disaster due to prolonged drought. According to the *Woreda*'s DPPO contact person, the trend in the number of emergency aid recipients is rising.

According to an assessment undertaken by the *Woreda* early warning committee, the number of people in need of emergency aid in 2010 increased from 22,505 (in 2009) to 24,184 due to the late initiation and early cessation of the main rainy season in 2009 and complete failure of the short rainy season in 2009. These erratic rainfall patterns and the concomitant severe drought led to scarcity of water, decline in pasture and crop productivity and ultimately resulted in food insecurity for pastoralist households.

Though the climate in the study area is very erratic and unpredictable, a deeper analysis of the local context reveals that the pastoralists-state relations resulting in denial of access to the Awash River is at the centre of the problem. As illustrated in the table below, the pastoral informants from the study villages list the different social aspects of vulnerability that have influenced their adaptation strategies over the years. However, they also mention the influence of climate-related forces such as drought on their routine practice of livestock mobility.

Table 1: Pair-wise ranking of major sources of livelihood insecurity

Pair-wise ranking	Decline in grazing area (1)	Increase in human population (2)	Decline in customary institutions (3)	Frequent and severe drought (4)	Conflict with other groups (5)
Decline in grazing area	X				
Increase in human population	1	X			
Decline in customary institutions	1	2	X		
Frequent and severe drought	1	4	4	X	
Conflict with other groups	1	2	3	4	X
Score	4	2	1	3	0
Rank	1st	3rd	4th	2nd	

Source: Compiled from the FGDs and participatory exercise (2010)

This also justifies my usage of the political ecology approach to understand the localized effects of climate change in the dry land context of Ethiopia. The side effect of both drought and loss of access is conflict, as the Karrayu pastoralists are obliged to travel to distant sites with their animals, and this causes conflicts with other groups along the way. However, the construction of irrigation canals and the diversion of the river to the villages have tremendously eased the problem of water shortage in the area.

Figure 11: Major trends, events and responses in Fentalle Woreda

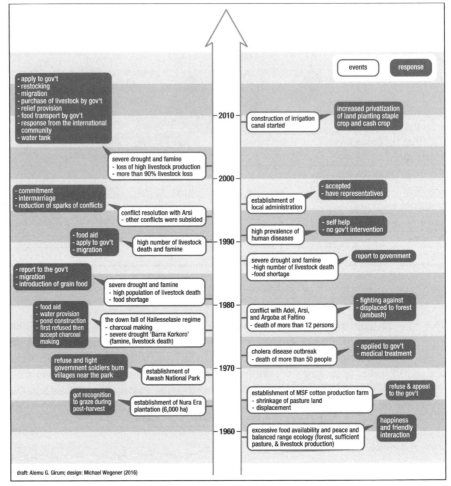

Source: Compiled from participatory time line mapping (2010)

The 1970s and 1980s droughts in the area and the associated loss of livestock caused great havoc which was compounded by the loss of key grazing sites. The Upper Awash Valley was also seriously hit by the 1973–1974 droughts, which had a national scope. During this period, both plant species and animals were affected, due to the extended dry periods. As a result, many valuable shrubs and grasses disappeared from the Karrayu rangeland. According to my key informants, a number of grass and tree species either disappeared from the vicinity or are fast dwindling. For instance, they mentioned a very important grass species, *Enteropogon macrostachyus* (*deremo* in the local language), that disappeared completely. This decline in both quality and quantity of the pasture land has hampered the ability of the pastoralists to deal with climate-related stresses within their habitual territory.

These empirical observations from Karrayu pastoral areas clearly support the perspective of hazard that locates it in the wider social setting. In this regard the 'critical realist perspective on hazards' captures the situation among the Karrayu very well. As Wisner *et al.* (2004) point out in their pressure-and-release model of disaster, risk is located in the interplay between certain non-human processes or events which interact to produce underlying social dynamics and unsafe conditions, such as those produced by large-scale development interventions in Upper Awash Valley. During my fieldwork in 2009, I observed that drought had claimed the lives of a large number of animals belonging to Karrayu pastoralists, especially goats. Recurrent droughts in the Karrayu community have resulted in the loss of a large number of animals, and this continues to be a challenge. However, the impact of drought in the past and the resultant vulnerability to famine is not solely due to the failure of rainfall and misuse of resources by the locals. Rather, the development-induced disconnection of the Karrayu pastoralists from their primary resource base and the associated loss of access to key grazing resources were responsible for compounding the impact of drought, which was further aggravated by diminishing natural resources and environmental degradation. Getting access to flooded grazing areas into which they traditionally move during the wet season is crucial to the Karrayu pastoral households if they are to survive and to exploit the area. When a vast tract of land close to the river is inaccessible for dry season grazing, a much larger area far away from the flood plains is rendered almost useless. The denial of access to these productive resources exacerbates the problem that the pastoralists already face, and which in turn has repercussions on the wider social and environmental situation of the Karrayu. This has resulted in the exposure of the Karrayu pastoralists to drought and famine through an orchestrated transition of their communal grazing land into private and state property over the last fifty to sixty years.

4.3.2 Pastureland degradation and encroachment of invasive plants

Ecological degradation and deterioration in the condition of the pasture land resources has also contributed to the process of livelihood vulnerability among the Karrayu. The focus group discussion participants mentioned three major factors that contributed to these changes. First, the mushrooming of small farms has resulted in private enclosure of the grazing areas, making traditional herding practices impossible. In addition, the associated introduction of leasing out of land has sparked competition among the herders themselves in the well-watered Karrayu areas. Thirdly, the newly introduced irrigation scheme that was meant to combine both livestock and farming practices has paved the way for the introduction of cash crop production in the area. In the field, I observed that a large tract of land has been allocated to cash crops without any proper consideration to the nature of the dry land and the pastoralists' style of social organization. In some villages, those who farmed in recent years have already started to complain about soil salinity. Another environment-related source of vulnerability is the increased encroachment of non-palatable species in the rangeland. Most of the bush and trees are used as browse in the pastoral areas of Fentalle *Woreda* and the neighbouring pastoral communities. The importance allotted to browse in natural grazing lands varies according to where the browse is available. Palatable grazing and browsing species are also being severely reduced as non-palatable species increasingly dominate. The bush encroachment reduces the grazing lands of the area and this leads the pastoralists to compete for open grasslands. This is one of the situations that forced the Karrayu, the Ittu and the Afar pastoralists to take their herds to the open grassland of the park. For example, *Acacia orfata* can extensively be seen in the northern and north-western parts of Metehara town, including the park. Unless bush encroachment is duly controlled, it would be a serious problem for the sustainable management of the rangelands. The reasons for the encroachment are many and include land degradation, overgrazing, drought, and unwise use of rangelands.

Photo 3: Degraded pasture in the Karrayu rangeland

Source: Author

According to a range condition assessment conducted by the regional government, the rangelands in the Metehara plain are severely degraded (OWWDE, 2010). According to this unpublished report, the grazing and browsing capacity of the Metehara plain has declined and less desirable plants have rapidly replaced the more palatable and nutrient species. Notable abuse of rangelands has occurred in low rainfall arid areas under inadequate systems of grazing management and where periodic droughts have hastened the effects of extensive grazing. Most of the plains area is severely degraded, wide margin areas are heavily grazed, and the soil is left bare of all vegetation. A vicious circle of natural resource deterioration-drought-starvation has been induced by rangeland degradation. The principal causes of rangeland degradation in the Fentalle district are drought, ovegrazing, expansion of cultivation, bush encroachment and soil erosion.

According to government rangeland experts, the indicators of degradation include undesirable and invader plants. The invader plants in this case are *Perthineum, Calotrops procera, Tribulis terristerhuis, Solanum species, Sida ovate, Cryptostegi grandiflora, Brucae antidysenterica, Datura stramineum, Amaranchu* and *Acharantus aspara*. Rangeland degradation needs to be controlled by checking the growth of these plants.

4.4 THE POST-1991 STATE IN PASTORAL SPACES: A LIABILITY OR AN ASSET?

As I have explained (see section 4.2.1), since the middle of the 20th century, the previous two regimes in Ethiopia embarked on an agenda of 'modernizing' the country through a heavy-handed centralized planning system. Pastoralists inhabiting the lowland peripheral areas of the country were excluded from participating in the 'development interventions' that directly affected their very existence, and were also excluded from any benefit that these modernization schemes brought to the central government. However, since 1991, Ethiopia has started to experiment with a new political order by putting in place an ethnic-based federal system. By examining the ongoing state-driven development interventions in Fentalle *Woreda*, I bring to light the various forms of state-pastoral interactions that in turn shape the arrangements and configurations of natural resources and livelihood activities.

4.4.1 Political reconfiguration and the recognition of a pastoral way of life

There have been recent changes concerning the pastoralist policy of the government. Following a reorganisation of government structures which took place in 2001, an inter-ministerial board was established within the Ministry of Federal Affairs (MoFA). This board was intended to serve as a secretariat for pastoral issues. While the establishment of this unit was a sign that the government was taking pastoral issues seriously, its institutional location could be interpreted as an effort to control the issue at a federal level. Furthermore, there is concern about the extent to which ministers with broad mandates will be able to give sufficient attention to pastoralist issues, as well as the extent to which the board will be able to coordinate the pastoralist units in different ministries. There is also a call for a separate pastoralist ministry or commission. This is currently being resisted by the federal government, mainly on the grounds that it is the responsibility of regional governments to oversee pastoral issues, and the technical weakness at this level should be addressed. After a national consultation workshop in 2002, a statement about pastoral development was issued, being the first official declaration of this sort ever issued by the Ethiopian government. The position expressed by this document was in favour of long-term voluntary settlement of the pastoral population.

Another recent change is within the parliament, where a number of new standing committees were established with the mandate to oversee government bodies. One of these is the Parliamentary Standing Committee on Pastoralist Affairs (PSC), established in mid-2002. The Pastoralist Communication Initiative (PCI) was subsequently formed and started to meet with pastoral parliamentarians, initially individually and then by region. International academics, Kenyan pastoralist leaders and others participated in further discussions with the pastoral parliamentarians about the establishment of a parliamentary group. The pastoral parliamentarians proposed a small sub-committee of eight people tasked with setting up a standing committee. The membership of this interim committee was decided at regional level and it was chaired by a Somali person. The interim committee embarked on the task of establishing a standing committee, with the support of the PCI. After a period of negotiation, the establishment of the PSC was allowed. Five particular factors contributed to the successful establishment of the PSC. Firstly, the increased political attention to pastoralist issues, within the context of broader political changes, and moves towards political pluralism, provided an open space for changes to occur. Secondly, a concurrent restructuring of parliament which reflected changes in the structure of government and the establishment of new ministries provided a specific political opportunity for structural change. Three new standing committees were created at this time. Thirdly, the support of the Speaker of the House was essential. He accepted from the outset that a critical constituency was historically marginalized and structures needed to be put in place to give greater voice to that constituency. Fourthly, the mobilization of ninety parliamentarians, including strong pressure.

After coming to power in 1991, the EPRDF formally institutionalized ethnicity by embracing a federal approach. The preamble of the 1995 constitution of Ethiopia emphasizes the "requirement of full respect of individual and people's fundamental freedoms and rights" (FDRE Constitution, 1995: Preamble). Compared to the historical marginalization of people residing in the distant lowland areas of the country, the new constitution of Ethiopia is generous in giving recognition to the social, political and economic values of "nations, nationalities and peoples of Ethiopia".

During the previous two regimes, the cultural and social practices of pastoralists have been degraded by using derogatory words such as 'zelan', which literally means wandering aimlessly, and their land was considered to be no man's land ready to be developed for other purposes. This situation was worsened during the military regime, as the pastoralists' way of life was seen as 'backward' and as a hindrance to the governments' modernization programmes.

However, the post-1991 period marked a significant departure from the previous regimes by granting constitutional recognition to the various ethnic groups residing in the marginal lowland parts of the country, such as the Karrayu and the Afar, among others. This ethnic-based constitutional recognition of people's right is also extended to the granting of rights of utilizing, managing and administering the natural resources and territories that they inhabit.

4.4.2 Transforming the nomads: rationality of state development projects

Although it has granted some form of self-government to historically marginalized groups in principle, the EPRDF has continued to play a centralizing role by reducing regional and local administrative divisions to the status of implementers of policies made in the centre (Clapham, 2009). This ambivalent role of the federal state and contrasts between the constitutional provision and practical realities are part of latent tensions between regions and the federal state, as highlighted in the case of the management of critical resources such as land and water. Despite the incorporation of the lowland peripheries into the Ethiopian empire in the late 19th century, state presence was minimal under the past regimes (Hagmann and Alemaya, 2008; Markakis, 2011). However, the EPRDF government has successfully reached the periphery, both through its ethnic-based administrative structure and in the form of development projects that are aimed at taming the wilderness and converting it into an economic resource. It has departed from its predecessors in reversing the old assumption about the periphery as 'marginal' in its contribution to the national economy. The periphery has become a viable economic resource in the form of providing ample arable land for private investors and government projects. In the following sub-sections I discuss the rationality of state development interventions in the lowland areas of the country and the approaches followed by the state in helping pastoralists to secure their livelihood and manage risks. State intrusion into the periphery during the post-1991 period is mainly due to the government's ambition to diversify the economy to foster development. These development projects in the pastoral areas of the country are in essence similar to previous top-down interventions in pastoral areas. They take the form of building dams to generate electricity and establish large-scale irrigation schemes. These interventions are justified by the government as a process of modernizing and transforming the people in the region.

4.4.2.1 The pastoral land issue and development in Ethiopia

Though pastoral rights have received recognition at the highest level through embodiment in the Federal Constitution (1994), in practice their protection is poor and formal land tenure systems have not yet been developed for the pastoral areas. Further, the weak capacity of the regional and lower level government administrations has limited the use of their power (Helland, 2006). The general policy statements enshrined in the constitution have been elaborated and specified through detailed rules and regulations stipulated under the federal and regional rural land laws. The federal and regional rural land laws and the detailed rules and regulations therein focus predominantly on governing an individual landholding system that gives less emphasis to communally held resources. In other words, the existing land laws of the state fall short of practical implementation and positive impacts in pastoral communities, as these laws essentially ignore the local contexts, customary tenure arrangements, and resource use and governance practices of the pastoral communities. Despite non-recognition in the state-backed formal land tenure policy, and rural land laws, pastoral communities in the country have been managing and using land and related rangeland resources for centuries based on a communal land tenure and governance system. The pastoralists have managed to sustain this system through a complex and well-structured web of customary institutions, rules, regulations and principles that underlie the use and management of pastoral land and related rangeland resources. The Ethiopian land tenure system has continuously failed to integrate and accommodate this time-tested communal land tenure and governance system of the pastoral communities. The predominant focus of land policy formulation and legislation in the country has always been on a farming-based land use system, essentially disregarding the details of communal land tenure, property rights arrangements and the underlying customary institutions among the pastoral communities. This has led to continuous deterioration of the role and authority of customary institutions in respect of communal land use. As a result, the pastoral communities continue living under an insecure land tenure system characterized by lack of legal protection and increasing loss of land use rights and access to rangeland resources, as non-pastoral land use systems continue to creep into the pastoral systems (Elias, 2008; Flintan, 2011).

4.4.2.2 Fentalle integrated irrigation project: participation and pastoral development

The Ethiopian government has devised plans to transform the economy of the country and the lifestyle of the people, particularly in the periphery, through large-scale development projects. In the process, there have been clashes between promises of ethnic federalism and government development programmes (Markakis 2011). The Five-Year Growth and Transformation Plan (GTP) (2005/06–2009/10) and the PASDEP (Plan for Accelerated and Sustained Development to End Poverty) focus on 'livestock development', water provision, forage development, irrigation and marketing (MOARD, 2010). Subsequently the regional Oromia government has developed its own programmes and projects in respect of the pastoralists and agro-pastoralists residing in the region. One of the strategies that have been devised by the regional government is the integrated area-based development approach that divides the entire region into three growth corridors. According to the document, the approaches followed are "problem driven and are aimed at socioeconomic transformation of certain affected areas in a sustainable manner". To this end the development corridor approach, which enables area-based development interve- ntions, was chosen as a strategy by the

Oromia regional government to meet set development objectives. The marginalized lowland parts of Oromia, which are hit by recurrent drought, are addressed in this programme. However, important factors such as securing access to land and resources are not mentioned.

Similar to the previous two regimes, which sought to change the 'backward' practices of pastoralists to more 'civilized' sedentary farming, the EPRDF aims to settle pastoralists, whose livelihoods are considered as not sustainable in the face of an ever-changing environment. Furthermore, the dominance of smallholder agriculture in government policy-making has facilitated smallholder expansion into areas previously used by pastoralists, in particular through sedentarization programmes, and a marginal lands narrative has been employed to question pastoralists' existing land use and justify large-scale investment leases. The government also aims to reduce poverty and build adaptive capacity in pastoral areas through sedentarization (FDRE, 2002).

The current government has developed mixed messages concerning development in pastoral areas: on the one hand the government highlights the contribution of pastoralism to livestock production (MOARD), identifying it as an area for further economic growth and linkages with non-pastoral economic systems (MOFA, 2008); and on the other hand it states that it anticipates that all pastoralists will be settled (FDRE, 2002; 2006; MOFA, 2002). The government argues that settlement is the only viable option to reduce poverty in pastoralist areas and that cultural transformation of pastoralism is a prerequisite for the sedentarization initiatives to be successful (FDRE, 2002). The government's agenda of changing the pastoralists' way of life by transforming their environment is also pushed by the Oromia regional government. Based on the government's assessment of the relative values of the different forms of land use and production systems, the Oromia region has crafted a five-year strategic plan, a water-centred growth corridor, which divides the entire region into three zones. As part of several projects, the Fentalle project is also meant to help the Karrayu pastoral groups residing in the area through an integrated project that combines crop-farming and livestock production. According to the Oromia Water Works expert who works in the project, they are planning to integrate crop farming with livestock production. He further commented that as part of the project they created a meticulous land use plan based on soil variety. They identified priority areas of intervention for the *Woreda*. In his view the diversion of the canal is an example of excellent engineering, using irrigation technology from Israel.

The idea behind the Fentalle integrated irrigation project is dictated by national discourses on pastoralists' development that favours a settled farming-based way of life over mobile pastoralism. Government officials usually mention that at the end of the day the government do not really appreciate pastoralists remaining as they are. The government wants to improve the livelihood of the pastoralists by creating job opportunities. This is further reflected in the various regional project documents. The overall logic in the document is to settle the pastoralists and they are expected to keep only small numbers of productive livestock. They will be provided with water and fodder and will be connected with the regional market. Following a highly modernist idea of development, the regional government is planning to organize some of the Karrayu households as milk producers and processor cooperatives. Through thin simplification of complex realities on the ground the project further envisages to connect Karrayu households with agro-processing industries that create job opportunities and their strategic location along the Addis Ababa-Djibouti highway is seen as a reason to transform the area into a regional economic hub. This is in line with what James Scott (1998) calls the "administrative ordering of nature and society". Such social engineering schemes are often a result of the state's need to simplify local practices and to make them legible for their bureaucracies, which has detrimental effects on local communities. Parallel to Scott's (1998) argument, the government's modernist project in pastoral areas has failed to take account of the pastoralists' practical and experiential knowledge, which is required in order to manage the complex environmental and social realities.

As part of the integrated irrigation project, livestock production has also received attention. However, rather than an extensive and mobile form of livestock production, the project intends to introduce a paddock system for effective animal distribution over grazing land. This is based on a highly rationalistic idea of management with precisely defined and rigid units. For instance, one of the problems of the integrated irrigation project planning process is that it does not consider the local variations in livestock holdings and simply divides the grazing land in the project area into paddock units for the promotion of uniform utilization of the vegetation, and for monitoring purposes. The land units identified and recommended for forage production are to be developed and used through the establishment of ecologically suitable grasses and legume species in respect of both adaptability and productivity. The recommended grasses and legumes are mixed varieties based on soil fertility, economic use and the interests of the pastoralists and farmers.

According to the Oromia Water Works and Design Enterprise (OWWDE, 2010), forage cultivation can be rain-fed depending on the location of the sites. However, to get maximum economic return from the land use plan, the use of irrigation to supplement rain-fed forage cultivation is advisable. With supplementary irrigation-based forage cultivation, the production of forage grasses increases more than three-fold and the livestock stocking rate per household, productivity and economic return, will also be tripled in the same manner (OWWDE, 2010). It is planned to use the water in the form of spate (furrow) irrigation in the sown area using gravity. Herbaceous legumes forage is grown in a relatively intensive way to improve dry-season nutrition of animals by improving the quality of standing hay and crop residues. Fodder trees should also be considered as one of the potential means of forage production. The ideal place for fodder trees and planting materials is along the buffer area of irrigation canals and around the homestead. Fodder trees like *Leucaena leucocephala, Sesbania sesban* and *Gliricidia sepium*, and planting materials such as *Sorghum sudanse* (Sudan grass) and *Pennisetum purpureum* (elephant grass) are recommended because of their adaptability and production potential. The land use plan for forage development at proposed new settlement sites in Fentalle *Woreda* is described in the following table.

Table 2: Area recommended for forage development user households

Project name	Site name	Area (Ha)	Proposed user HH's (0.15 Ha/HH)	Current land use
Fentalle	Amuma	193	1,287	Open communal grazing land
	Bedenota	107	713	Open communal grazing land
	Sala	301	2,007	Open communal grazing land
	Areda Ilma Sebeka	190	1,267	Open communal grazing land
	Total	**791**	**5,273**	

Source: Based on data from Oromia Water Works and Design Enterprise (2010)

This is what Scott (1998) has termed as the states' idea of development that is based on uniformity and undermines local knowledge. According to this centralized planning process, the proposed settlement sites in Fentalle *Woreda* will be provided with at least one paddock system. However, the paddock systems assigned to each settlement site vary in size depending on the land available, production potentiality and suitability for fencing and management operations. When it comes to putting the political rhetoric into practice there is a huge gap. For instance, in contrast to the much discussed participatory approach to development, hierarchical structures persist in state-pastoralist relationships with regard to implementing development programmes. Rather than providing political spaces for citizens to contest and negotiate discourses and practices of development, planning has become a technocratic and top-down process. In other words, despite the rhetoric, the new political order has demonstrated a retreat to the hegemony of the centre, with strong state intervention in local affairs, specifically in the resource-rich pastoral areas, particularly since the early 2000s. During my field research, a few government officials at the *Woreda* level expressed their concern about the top-down nature of the ongoing irrigation project. In the words of one official: "The higher officials did not consult us on the project. They simply came and told us to mobilize the people. They simply burden us with the assignment of convincing the Karrayu" (Metehara town, 2010). Other *Woreda* officals argued that the project lacked proper planning, and had resulted in the wastage of resources. Furthermore, they criticized the parallel implementation of contradictory projects in the *Woreda*. The *Woreda* experts illustrated the top-down nature of the project with the image of a hot potato that is passed from hand to hand, from the higher level down to the zone and the *Woreda*. In the end, it is the *Woreda* officials who face the consequences.

This view of the government officials is widely shared among the Karrayu community. While the government officials encourage the local people to engage in debates concerning issues that are important to their communities, as a sign of democratization, the fact that those who make critical speeches during local meetings are prevented from speaking can be seen as a strategy of suppressing critical voices. This compromising of political rights through development interventions is evidence of the EPRDF's prioritization of economic development over democratic rights. During the inception of the Fentalle integrated project, for example, various experts came together to air their views and discuss possible intervention pathways. However, according to one official who had been part of that discussion, not everyone was given an equal chance. Rather, people like him, who recommended a livestock-led development intervention in line with the nature of the environment and the Karrayu people's long-time engagement with livestock keeping, were shrugged off by higher officials. Thus, the emphasis on sedentarization and transformation of pastoral lifestyles and cultures reflects the Ethiopian government's top-down approach, even though on paper the government claims that its strategies are participatory (Davies and Bennett, 2007; Dyer, 2013).

4.5 IN-MIGRATION AND INCREASED POPULATION PRESSURE

Another important social source of vulnerability, which has influenced livelihoods in the Karrayu community and their natural resource base, is an increase in population pressure on Karrayu land. Both oral history and documented evidence indicate that Fentalle *Woreda* enjoyed a good resource base until around the end of the 1960s. The combination of ecological variables (availability of different streams, salinity levels and pace of water flow) provided a productive ecosystem and habitat for numerous species of flora and fauna. The environmental history of the Karrayu land shows that the state of its ecosystem with respect to health and resource sustainability acted as a crucial driver for the emergence of communal resource use.

On top of this, a reasonably low population density was an additional factor that helped maintain the resource base. Though it is difficult to find any written record of the population size of a pastoralist group like that of the Karrayu some seven decades ago, the elders in the community estimate that in the 1940s the total population of herders in Fentalle *Woreda* of Upper Awash Valley was about three to four times less than it is today. Harris (1844:209 as cited by Kloos, 1982) estimated the Karrayu population to be around 5,000 when he visited the area in the mid-19th century. From this population figure one can easily deduce that the small population had enabled the Karrayu herders to function cohesively enough to form village groups that could craft and handle the rules and norms of resource use and resolve resource conflicts. However, after a century and half, the population in the rural part of the Fentalle district was 61,708 (CSA, 2008).

Figure 12: Trends in population number of the Karrayu since 1968

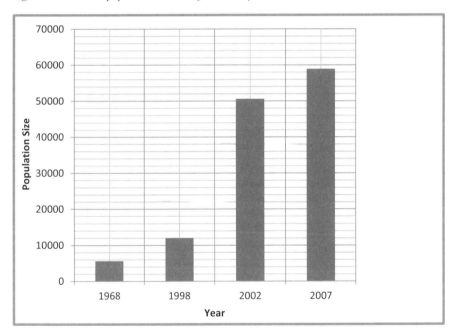

Source: Data based on Muderis (1998) and Fentalle Woreda Agriculture Office

Another important demographic factor is the movement of non-locals into Karrayu territory, which has put a lot of stress on the limited resources that have been left after encroachment by the large-scale commercial farms and conservation measures. Particularly important here was the immigration of the Ittu which increased in magnitude in the 1950s and continued throughout the 1950s and 1960s as a result of war with the Issa Somali over resource use and ownership in their ancestral land.

After the migration, some Ittu settled alone in villages like Benti, and others lived together with the Basso Karrayu in Kobboo, Degha Hedhu, Galcha, and Moogassa (Ayalew, 2001). The first three are located in the foothills of Mount Fentalle, and the fourth near the sugar cane plantation and the park. The relations between the newcomers and the hosts were good from the very beginning, since the migration was conducted in agreement with the Karrayu. The friendship and solidarity between the Ittu and Karrayu that characterized the pre-migration period were further strengthened after the migration through marriage, reciprocal exchange of stocks and gifts, and cooperation in the payment of blood compensation in particular and pastoral life in general.

Figure 13: Migration routes of the Ittu to Karrayu land

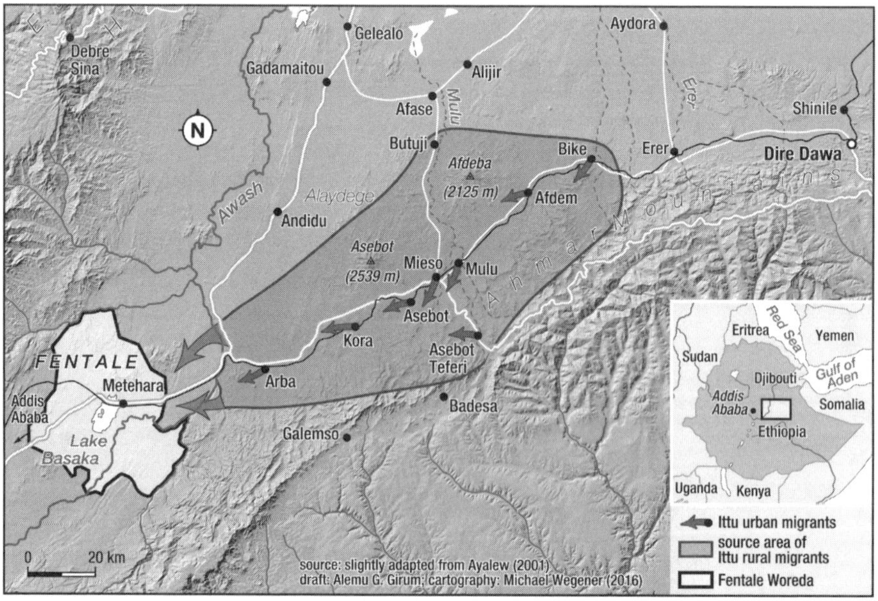

Drought triggered the migration of the Ittu only in 1984/85. Otherwise, what made the migration of the Ittu of this period different from the previous ones was that until 1984/85, it was only the pastoralists who migrated to Fentalle. But this time, the absence of rain for two consecutive years and the subsequent drought and famine in west Hararge also forced the agro-pastoralist Ittu to migrate to the Karrayu territory in Fentalle *Woreda*.

4.6 SUMMARY: MULTIPLE SOURCES OF LIVELIHOOD INSECURITY

This exploratory chapter has presented the broader trajectories of livelihood insecurity that are rooted at the intersections of pastoralists-state-environment interactions. Its purpose was to situate Karrayu pastoralists' risk management and livelihood practices within broader historical contexts. Accordingly, it is a reflection on the factors and processes that have been reducing and limiting the ability of the pastoralists to pursue their livelihood practices in the semi-arid environment, and to make clear that most of the causes of vulnerability to climate stress have little or nothing to do with climate, for almost all of them stem from capitalist and modernist notions of development.

As stated in Chapter Three, the political ecology approach sees pastoralists as culturally adaptive agents to the limitations of nature but whose basic relationship to nature has been negatively affected and eroded by the working of structural processes. By emphasizing on the political-economy systems of the various Ethiopian regimes, this chapter highlights the relevance of contextualized and multilevel analysis of livelihood insecurity in upper Awash valley. Accordingly, unlike the dominant adaptation discourse that focuses on climatic stimuli as the major driving force behind livelihood insecurity of the pastoralists, this chapter approached and contextualized problems of pastoralists' livelihood in the broader political and economic trajectories of change. This helps us to better understand the influence of access to and control over resources and social relations of production on the risk management and livelihood security of the locals.

The political ecology approach sees the problem of the Karrayu as arising from their integration into the wider socio-political system and the subsequent policy interventions by successive Ethiopian governments which resulted in the loss of traditional pastoral autonomy. This excluded them from making decisions concerning their own affairs. Pastoralists were denied their customary rights of tenure and access to land. This transfer of ownership of pastoralists' land to the state led to the expropriation of their best grazing grounds and water points.

The major large-scale development interventions after the Second World War in the river valleys of Ethiopia took place in the Awash Valley, and a sizable number in 'Karrayuland'. These developments, in turn, caused various environmental problems. The Karrayu pastoralists were confined to an ever-shrinking area if pastureland, which resulted in the drastic degradation of natural resources. The development schemes led to land alienation in the form of built-up areas (Metehara and Addis Ketema towns), agro-industrial units (Nura Era State Farm, Metehara Sugar Factory, Abadir Farm), conservation schemes (Awash National Park), and an expanding lake (Lake Beseqa). Each of these projects has contributed to a decline in the land holdings of the Karrayu, and has failed to include them in development decisions and to effectively compensate them for their losses. In this chapter I have shown that it is the loss of access to their previous wet and dry season spaces of adaptation that has shaped the vulnerability of the local pastoralists to climate-related extremes, such as drought. These historical processes have serious implications for the adaptation strategies followed by the Karrayu pastoralists today. The major state-sponsored ventures in the dry and wet season grazing sites of the local pastoralists are illustrative of the imbalance of power between the state and the locals, which, compounded with drought, has resulted in their increased vulnerability.

The conventional top-down approach in climate change adaptation research puts emphasis on anthropogenic factors that lead to the deterioration of adaptation and insecurity of livelihoods. By contrast, in this chapter I have argued for starting-point vulnerability, which puts the climate change adaptation debate into its proper social and political context. The resource access problem faced by the Karrayu cannot be attributed solely to the direct loss of grazing land due to the establishment of commercial irrigated farms and the Awash National Park, but also to population pressure. The problem of pasture land has been exacerbated by the demographic pressure that came about with the continuous encroachment and eventual settlement of the Ittu on the traditional land of the Karrayu. This confirms the fact that local-level vulnerability, like that experienced by the Karrayu pastoralists, is contributed to by processes that have broad-scale resonance and origins, and are outside the direct control of individuals, households and communities (see the figure below for a summary of the broader context of livelihood insecurity). The Karrayu were entirely dependent on livestock rearing for centuries. As a result of various sources of vulnerability, however, they have begun to combine stock keeping with subsidiary economic activities like cultivation, in order to deal with the various sources of vulnerability, as will be discussed in the subsequent chapters. In line with the foregoing, in Chapters Five and Six I will explore how the Karrayu pastoralists have used their agency to handle vulnerability arising from both social and climatic sources.

Figure 14: Summary of the context of livelihood insecurity

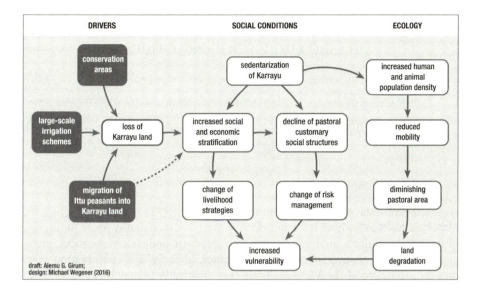

5. ENVIRONMENTAL TRANSFORMATION AND LIVESTOCK-BASED LIVELIHOOD PRACTICES

"There is no life without our livestock. We respect and value our cattle and prepare their kraal in front of the house ...our status used to be measured on how many cattle we owned."

[Karrayu pastoralist, June 2010]

5.1 INTRODUCTION

As discussed in the previous chapter, the various Ethiopian regimes conceived the Metehara plain as nothing but unproductive marginal land that is ready for the use of modern agriculture. In doing so, they disrupted pastoralists-environment interactions. With the coming of large-scale commercial farms and conservation schemes the landscape of the Metehara plain was drastically changed and the ecology has been transformed. These development interventions and the associated environmental transformation denied pastoralists access to crucial natural resources – pasture and water. Prior to the development interventions, the major livelihood activity practised by the pastoralists was extensive livestock production. However, the denial of access to important resources has seriously jeopardized their cattle-based livelihood practices. In this chapter I discuss the social disruptions that have resulted from the modernization processes that occurred in the valley from the mid-1960s and the continuity of livestock-based pastoral risk management and livelihood practices in the face of changes in the environment.

5.2 ENVIRONMENTAL TRANSFORMATION AND SOCIAL DISORGANIZATION

The bio-physical environment in which the pastoralists operate influences their forms of organization and resource utilization, as well as the adaptation activities in which households engage. As has been discussed in Chapter Four, the environmental transformation brought about by the state interventions has resulted in the disruption of the Karrayu mode of relations with their environment and their social organization which guided their access to, utilization and management of resources for centuries. Prior to these development interventions and the resultant environmental changes, cattle rearing were central to their livelihood and cultural survival. However, the disruption in pastoralists-environment interactions and the associated loss of access to key grazing land made the practice of cattle-oriented pastoralism almost impossible. The immediate consequence of the denial of access to resources was disruption in their mobility patterns and social disorganization. In the following sections I discuss the social and institutional ramifications of distortion in pastoralists-environment relations.

5.2.1 Distortion of settlement and mobility patterns

Mobility is one of the major hallmarks that characterize communities engaged in pastoral livelihood practices. Besides serving as a mechanism to tackle resource scarcity and associated environmental sources or risks, it is also an important practice through which the alleged ecological degradation is avoided. Mobility makes possible the optimal exploitation of resources scattered in time and space and maintains the natural process of adjustment between human population, livestock and natural resources. Before the wide-ranging changes that drastically affected their environment, the Karrayu pastoralists were considered more sedentary than nomadic, as "they seldom moved more than fifty kilometres away from their watering places" (Haberson, 1978: 478–479). They travelled between their permanent neighbourhoods (*onnetesso*) and the migration area (*beke dheda*) (Ayalew, 2001). They used to be seasonally semi-nomadic, moving all their possessions to two distinctly separate seasonal residence sites. The history of pastoral mobility in Karrayu, over long periods, has been one of increasing distances of migration, diminished access to resources, and smaller social units of resource users. Here I discuss briefly the historical patterns of mobility of Karrayu households and how they have changed over the years.

In earlier decades, Karrayu pastoralists had a simpler pattern of seasonal migration, freely shifting back and forth between higher grounds for wet season pasture and the Metehara Plain and banks of the Awash River for dry season pasture. Traditionally, the pattern of mobility was highly orderly and regular, as well as limited in terms of distance. The occupation of sizable tracts of Karrayu pastoral land and major watering points (*melkaa*) by concession farms has, however, distorted this pattern. The shrinkage of the resource base that resulted from large scale development interventions affected the patterns of mobility which became less regular, causing disruptions in the land use pattern of the Karrayu and forcing them to travel longer distances in search of pasture and water. The Karrayu pattern of transhumance and system of land use had three forms prior to the middle of the 20th century. The Karrayu refer to these as *Onna Ganna, Onna Birra,* and *Onna Bonna.*

Table 3: Major seasons and mobility destinations in Fentalle Woreda

Months	Weather conditions	Local name	Places and destinations
January-May	Long dry season	Onnaa Bonna	Northern banks of the Awash river, around Gaalchaa Village
June-August	Long rainy seson	Onnaa Gannaa	Mount Fentalle area, north of the crater
September-December	Autumn season	Onnaa Birra	Southern foothills of Mount Fentalle
In February 2–3 days rainfall		Furmataa	

Source: Compiled from key informant interviews (2010)

The above table summarizes the seasonal mobility pattern of the Karrayu pastoralists within the Metehara vicinity of Fentalle *Woreda* by maximizing resource variations across space and time prior to development interventions that led to its disruption.

The permanent settlements where the Karrayu pastoralists used to spend the wet season between the months of June and early September are locally termed as *Onna Ganna*. These traditional settlement sites include Dhebiti, Haro Kerssa, Balchi and Harrokersa. Most of these settlement sites are located at Mount Fentalle, north of the crater. On the other hand, the settlement sites where the Karrayu herdsmen spend the period of time from late September until about the end of December is called *Onna Birra* (autumn season). During this period, the Karrayu herdsmen wait until the Awash River, which floods its banks during the wet season, has retreated in the autumn months and the flood plain pasture is ready for grazing their animals.

In line with previous studies in the area (Jacobs and Schroeder, 1993; Ayalew, 2001), the Karrayu elders with whom I discussed the issues of environmental transformation and mobility reiterated that they used to cover minimum distances in the vicinity of Fentalle *Woreda*. But today this has changed, because the increased population pressure in that part of the Karrayu lands left by the commercial farms has resulted in pressure on the limited rangeland resources. The high human population pressure is also attributed to the migration of the Ittu, that has gradually reduced the area available to Karrayu pastoralists and their herds. According to my informant Hawas Jillo, the rapid increase in population size was followed by shrinkage of grazing land, for a large part of it served as settlement area. This forced the Karrayu to graze their animals in a limited area over and over again. Because of the growing population, there was no surplus grazing land available to leave fallow. As is well known, surplus grazing land allows pastoralists to move to a different place and allow the pasture to regenerate. However, the increase in population has limited the grazing land, resulting in a scarcity of grazing and browsing resources for livestock, and has brought about intensive pressure on the rangeland resources. The continuous immigration has also led to an expansion of herds far beyond the maximum animal population that the land can support. This requires overuse of the vegetation within the limited areas of settlement.

As a result, overgrazing is common. It results from the imbalance between the number of livestock and the available grazing land or from shortage of grazing land. As the number of livestock increases due to population pressure, it restricts mobility and flexibility in time and space. This has destroyed forest and vegetation cover and reduced rangeland and livestock productivity and triggered sustained environmental deterioration in permanent settlement sites. One of my informants from Illala village, Nura Qumbi, expressed this gradual destruction of the fragile environment as follows: "Trees are cut now extensively in order to meet the increasing demand for wood, for making charcoal for sale and establishing new settlements. In contrast, the practice of replanting does not exist among us" (Nura Qumbi: May 2010).

Thus, in Fentalle, environmental degradation has come to be associated with the disruption of pastoralists-environment interactions and the concomitant increasing demand for scarce resources. Moreover, as resource access is restricted by the various historical forces explained in Chapter Four, and pastoral productivity declines with population growth, poverty-stricken pastoralists have to exploit natural resources in order to subsist. That is, with the decrease in livestock holdings, the number of pastoralists involved in firewood selling and charcoal-making activities has increased. This, in turn, has resulted in deforestation. Deforestation also occurs due to the high demand for firewood and construction materials. Furthermore, manure is not a significant source of fuel among the Karrayu traditionally. As such, the rapid population growth, with the coming of more and more Ittu, creates a huge strain on the region's natural resources and leads to destruction and fragmentation of resources.

The 'tragedy of the commons' sees the overexploitation of natural resources as rooted in the system of property. But the tragedy in Karrayu land is not due to the common use of natural resources, as claimed by theorists of the tragedy of the commons, who argue that rangeland degradation comes from the common use of, and free access to, resources. On the contrary, the pastoralists used to regulate the extent of resource use and conserve resources through various institutions. The real tragedy comes with external interventions, including the migration of populations and the collapse of institutions which manage common resource use. The political ecology model also criticizes the assumption of the tragedy of the commons that associates the overuse of resources with collective ownership of the rangeland. The political ecology approach sees this as ignoring the role of other Socio-economic systems. My informants upheld that overgrazing and deforestation, caused by overuse of resources due to overpopulation, and heavy economic dependence on them to cope with declining means of subsistence, have a significant impact on the environment and cause drought. As already stated the widespread practice of felling trees and degrading the soil has increased the risk of drought and made it recurrent in the study area. Thus, environmental degradation has its own impact on climatic change. It causes shortage of rainfall and increase of temperatures, with a consequent high rate of evaporation, so that surface water and ponds dry up. An informant from Giddarra village, Abba *gada* Kawoo, explained the reasons for the gradual weakening of their subsistence economy in the face of recurrent drought, by saying that the ability of the Karrayu livestock production system to resist the vagaries of rainfall has been seriously hampered by the over-population of the Karrayu land. He cited the loss of dry period grazing reserves to the irrigation scheme, which has affected their resistance to drought since it deprived them of access to dry season pasture and water. He added that drought exposed the animals to various diseases since their ability to resist disease was weakened. Consequently, vast numbers of animals perished from lack of forage and shortage of water. Under these circumstances, vaccination schemes cannot help to reduce livestock mortality. Drought also damages the crops that are cultivated by farmers who practise rain-fed agriculture besides livestock keeping. He concluded by saying, "We refused to sustain our environment, of course forced by external pressures, and as a result, our environment is refusing to sustain us." The environment has been affected due to unprecedented impacts arising from various sources.

5.2.2 Changes in the social fabric

Many of the structural sources of vulnerability discussed above have resulted in social change within the Karrayu community. This has seriously influenced the traditional pastoral mechanisms that make the Karrayu pastoralists resilient in the face of environmental uncertainty. The outcome is that the ability of Karrayu pastoralists to effectively deal with climate stress events or situations that may occur is lessened. It is important to emphasize that social change may also have beneficial outcomes for dealing with climate stress. However, during my field research I identified social change as the primary cause of increasing vulnerability. As a result of social change, the activities, processes and systems that reduce risks posed by climate variability and extremes are less woven into the fabric of everyday life and pastoral livelihoods than in the past two decades.

Four to five decades ago, everyday life and livelihoods were constructed and maintained in ways that accounted for climate stress based on generations of experience of living with environmental uncertainty. This was not necessarily purposeful or conscious, but merely part of the yearly, monthly and daily way of doing things. Participants in the focus group discussions felt that this is largely no longer the case, meaning that dealing with climate stress is less based on self-sufficiency and more uncertain. The reasons are embedded in processes of rapid social change that are associated with distant structural forces of vulnerability mentioned above.

Broadly speaking, the manifestations of social change mentioned by my informants in the Karrayu pastoral community can be divided into two kinds: 1) general socio-cultural changes, and 2) changes in pastoral practices. These two aspects are intertwined and were discussed by participants as priority concerns in the community, irrespective of climate stress. This was the strongest theme that emerged from the group discussions during the study. The participants clearly mentioned overarching changes in the social and cultural fabric of the community, including: knowledge, worldview, norms, values, belief systems, traditions, social relations and social organization. Socio-cultural change alters the social apparatus and livelihood systems holding traditional vulnerability reduction tools in place. In basic terms, this is because traditional vulnerability reduction mechanisms are embedded in Socio-cultural traditional knowledge, worldviews and values shaping livelihood systems.

5.2.3 Disruption in customary resource management institutions

In swiftly transforming environments and increased integration of the pastoralists into the national economy, the possibility to reanimate customary systems to govern pastoralists' resources is believed to be questionable (Lane and Moorehead, 1995). As has been generally testified, external intervention that alienated key land and water resources in the arid and semi-arid areas has put an immense pressure on the pastoral system and disrupted the livelihood of pastoralists (Ayalew, 2001; Getachew, 2001; Tache and Oba, 2009). Insecurity of losing land is ever high in the present situation of globalization and increased involvement of private investors in the once remote and distant areas of the rangelands.

Prior to the arrival of the large-scale development interventions, the Karrayu pastoralists used to rely on their well-organized customary institutions to manage their natural resources. The Karrayu territory that they call "Karrayu land", and all the important resources located within this territory, used to be governed by traditional authorities. The Karrayu used to divide their territories into various grazing zones (see section 5.2.1 above) and sub-territories, locally called *dheeda*. Within the Karrayu community, land was hold communally and the customary law didn't recognize exclusive ownership of any land or resource unit by any group or individual. According to my key informants, Karrayu land (*biyyaa Karrayu*) was owned, managed and defended collectively.

The traditional Karrayu methods of natural resource management and land administration are flexible and usually open to discussion with members of a particular grazing zone (*dhedaa*) under the leadership of the head of the grazing territory (*abaa dhedaa*). Under the leadership of the *abbaa dhedaa*, resources used to be governed and managed based on a set of principles, such as collective ownership, that respects the environment and protects it from overuse and degradation, and regulates access to key resources through sanctioning of free-riding, prohibiting spontaneous settlement and accommodating outsiders on the basis of reciprocity.

However, over time the customary resource governance structures of the Karrayu community have become almost dysfunctional, due to changes in ecology and land use following the state-sponsored development interventions in the valley. There are various exogenous and endogenous factors that have led to the further dwindling of customary institutions in the Karrayu community. The evidence from the field shows that population pressure, ecological disruption and the intrusion of non-Karrayu into Fentalle *Woreda* have together affected the structure and smooth functioning of the customary institutions and created tensions and conflicts among various groups. For instance, the spread of Islam in Fentalle *Woreda* is mentioned as one reason for the decline of customary methods of managing natural resources. The arrival and spread of Islam among the Karrayu pastoralists is associated with the migration and settlement of the Ittu from west Hararghe. Here it is important to stress that the migration of the Ittu has played a crucial role in the spread of Islam in Karrayu territory. But even though the influx and settlement of the Ittu in Fentalle *Woreda* is the major factor in the spread of Islam[1], sedentarization and the loss of traditional sacred places to the large-scale development interventions along the Awash River (see section 4.2.1) have also played a role in the spread of Islam and the consequent socio-cultural changes among the Karrayu.

Another important channel for the spread of Islam among the Karrayu pastoralists[2] is through inter-marriage. This requires conversion of the Waaqeffataa Karrayu to Islam, as the Muslims do not accept the marriage alliance without it. Using marriage and various social relations, the Ittu try to further strengthen their religion by converting the remaining followers of the traditional

---

1   The Ittu are committed to converting the Karryyu to Islam as a way of establishing closer social relations that would allow them continuous use of resources. Many of the migrant Ittu had religious teachers among them.
2   The traditional religion of the Karrayu like many other Oromos is called Waqqeffannaa and one who believes in the Oromo traditional religion is called Waaqeffataa.

religion. The Karrayu also show a preference for Islam[3] as against Orthodox Christianity, which is considered by them as the religion of those who have taken away their land through large-scale development interventions.

With the spread of Islam among the Karrayu pastoralists, some of their traditional customary social institutions are also disappearing. Here, I am not arguing that tradition is defiant of change. Traditions undergo transformation in order to go along with or adapt to changing conditions. However, the changes among the pastoral Karrayu in upper Awash Valley have been shaped by outside forces. The changes observed are endangering the continued existence of the Karrayu pastoralists. This has implications for the social fabric of the Karrayu, which is closely connected with the natural environment. For instance, the wider social ties created through the tradition of bride wealth payment are diminishing among the Karrayu with their conversion to Islam. Many of the converts and the Sheiks oppose the tradition of bride wealth payments by referring to the Sharia law. Even those Muslims who still expect the livestock payment do not require as many head of cattle as before. These pressures contribute to weakening the established pastoral practice of bride wealth payment that formerly served the purpose of creating cohesive social ties among clans.

Herders underline that due to the increase in population pressure and frequent drought it is not possible to return to the status quo. However, they still believe that the government is not giving proper attention to their customary ways of managing resources and continues to deny their rights to these resources through projects such as the expansion of the Awash National Park, irrigation development and certification of private land holdings. According to my informants, in some cases the certification of private holdings has led to conflicts within the community and created a sense of insecurity due to the loss of uncertified communal lands.

Another important aspect of the changing social and cultural fabric of the Karrayu pastoralists is the decline of their *gada* system. The *gada* institution that was crucial in orienting the social life of the Karrayu pastoralists has gradually become weaker. This institution has come under pressure with the continuing encroachment of state authority and the gradual loss of autonomy associated with political and economic marginalization. The coming into existence of government institutions such as *kebeles* has contributed to the decline in the importance of these traditional institutions (Ayalew, 2001). The expansion of Islam associated

---

3   The Islamization process among the Karrayu pastoralists has been supported by the Africa Muslims Agency (AMA). This organization has supported the construction of mosques in the Karrayu villages. It also provides financial support to those who visit Mecca on a pilgrimage and assume the title of 'Hajii'.

with the migration of the Ittu to Karrayu land has also weakened the role played by the *gada* institution among the Karrayu.[4] The functioning of the *gada* institution, which is related to the traditional religion of the Karrayu, is affected by conversion to Islam, as some of the political leaders (*abba bokku*) and the spiritual leaders (*qaallu*) have also converted to Islam and become Hajis. For instance, during focus group discussions in Giddara village, the participants mentioned the loss of societal orientation and leadership in the following manner:

> "Whom are we going to consult when it comes to our mobility? No one! Who is going to bless our land? No one! Most of our elders are now 'Hajis'. So many of them don't come to the celebration of *gada* anymore...because of that the young herders are doing what they want."

These processes of change affect not only the social fabric of the Karrayu pastoralists but also their economic situation. The gradual erosion of traditional institutions also jeopardizes the traditional mechanisms of managing risk and securing livelihoods, as these institutions were central to the proper handling of natural resources such as forest, pasture and water on the wider scale of the Karrayu territory. However, during my field research I found that the migration of the Ittu to Karrayu land has also influenced the Karrayu in a positive manner, such as the transfer of skills and knowledge in respect of farming practices, as will be discussed in Chapter Six.

As the Karrayu pastoralists' beliefs and worldviews change, so too does the social and cultural internalisation of risk management and livelihood practices. The management of risk that was designed for the arid and semi-arid environment that the Karrayu inhabit has become increasingly separate from general livelihood activities. The changing social apparatus of vulnerability reduction has important implications for adaptation to increasing environmental uncertainty with regard to climate change. Participants specified that Socio-cultural change is reducing the ability to apply, accumulate, transmit and adapt traditional knowledge and practices relating to environmental uncertainty. Socio-cultural change, as a structural force, is causing the traditional pastoral practices to be eroded. These processes of social change have resulted in the development of new practices among the Karrayu. The practical-evaluative aspect of their agency has led to the introduction of new mechanisms for managing risk and securing livelihood.

---

4   Even though some Karrayu who have converted to Islam still take part in the *gada* ceremonies, those converts who follow the *wabiy* sect strictly avoid them.

5.3 PASTORALISTS' AGENCY AND RESPONSE TO CHANGE

The ability to deal with changes in ecological conditions and the resultant environmental transformation arises from cultural, social and livelihood practices embedded in the processes of everyday livelihood and risk management practices. These practices are not necessarily consciously undertaken to reduce vulnerability to ecological changes per se, but often form the foundation of resilience. When research is focused on specific climate stresses and their impacts, as in the dominant impact-led adaptation research (Gaillard, 2010), the awareness of knowledge and strategies for managing risk and securing livelihoods may be restricted to those which are directly or obviously linked to specific climate stresses. However, while conducting field research and observing the Karrayu pastoralists' everyday livelihood practices, it is very challenging and difficult to distinguish decision-making activities tailored towards climatic stress from those due to other social factors. In the following section, I explain the processes of decision making in pastoral strategies with an emphasis on mobility so as to illustrate the concept of routinized pastoral agency within a changing context of vulnerability.

The processes of decision making by the Karrayu at a household level are connected with the broader social, economic, political and ecological environments. Within these broader environments, people undertake decisions, using the resources at their disposal and according to their knowledge of the various alternatives available, to come up with an outcome that provides the best possible outcome to their livelihood practices (Ensminger, 1996). In their everyday livelihood and risk management practices, Karrayu pastoralists exercise routinized practices such as the practice of taking out the small stock after the bigger animals have left the homestead. Other household practices such as determining who is going to take the bigger livestock to the distant pasture watering points and who is going to handle the small livestock in the nearby field, whether or not some small animals will be sold on the coming market day, or what kind of livestock will be given as a gift to a relative who is going to get married, takes more consideration.

Pastoral communities are depicted as homogeneous groups that take decisions unanimously. However, like any other decision scenario that brings different actors together, the involvement of many pastoral herders in decision making process could lead to contentious outcomes. This depends on the resources that each individual Karrayu pastoralist commands and can bring into the bargaining platform, and the agency of the pastoralists mediated by some cultural values. For instance, in my case study households, even though the women play crucial role in supporting their family, their role and agency are not fully recognized because of some culturally accepted values that depict the men as the provider of 'livelihood security' to the household. The same could also apply to young people who are in many cases has to follow the decisions taken by the elders on their behalf. These traditional structural forces that are exercised along gender and generational lines are usually unconsciously rooted in societal arrangements and influence the agency of individuals in decision making processes. However, this does not necessarily mean that individuals are always prisoners of culturally imposed structural barriers; individuals could take decisions against the existing societal arrangements as signs of resistance. Here, the consideration of various social factors such as seniority and avoiding curse are very relevant for the individual decision maker as the climatic factors are.

Karrayu pastoralists draw on available resources, and consider options these resources provide in dealing with the broader social and ecological contexts in which they operate. The resources and agency of Karrayu pastoral actors are influenced and mediated by traditional norms and values that have been in place for long time. This is in line with the conceptualization of the connections between individual actors' strategies and sedimented practices explicated in chapter two. For example, an individual Karrayu herder, using his agency and resources that he controls, could make decisions that will have consequences which in turn lead to new opportunities and/or constraints to his livelihood and risk management practices. In processes of decisions that involve several herders with their own agency and resources, the outcomes depend on the power differentials of the actors and how options are negotiated among the actors involved in the processes. As the number of actors participating in decision making processes increases, ultimately the resources, world views and knowledge that they bring into the negotiation platform increase, leading to increase in options that require more negotiations. In case of Karrayu herders, however, the negotiations work within some social frames organized along criteria such as gender and age that favour some individuals over others in the process of decision making.

Gendered and age-based agency among the Karrayu pastoralists

Mobility is conceptualized as a situated strategy to make use of geographically dispersed resources. Studying mobility in this way requires examining what enables pastoralists to be mobile and, conversely, what constrains their mobility. Clearly this also requires going beyond the household. Pastoral mobility is a scaled strategy, in that it is rooted not only in inter- and intra-household decision-making processes, but also within a broader political economy that organizes, assigns, and controls access to key resources. By viewing mobility in this way, it is possible to demonstrate how outcomes accrue to differentially situated actors in seemingly homogenous community. In the case of the Karrayu, it became clear in the course of my fieldwork that, depending on the ways in which they are socially positioned, individual herders are either constrained or enabled in the authority they have over mobility decisions. Concerning mobility decisions, for instance, it is accepted by the Karrayu households that the decision is the domain of the household head and sons only. Karrayu women have some control over small stock in consultation with their husbands. But the various customary organizations among the Karrayu, like the *gada* and *salfaa*, which guide overall decision-making on mobility, do not include women in their membership. All in all, the role of women concerning decision-making is limited to matters related to the permanent settlement areas, and they have little influence on mobility. This also tells us something about the differential and increased vulnerability of women in times of severe drought, as they have to stay at the settlement sites with the small children and the old people.

Pastoral livelihood and risk management activities among the Karrayu are differentiated on the basis of gender and age. The division of pastoral labour around livestock rearing is arranged and supervised by the head of the household who has the roles and responsibilities of managing the livestock. In the Karrayu pastoral community, as the head of the household and experienced herder, the *abba warra* (head of the household) makes decisions about and takes responsibility for the well-being of the livestock, on which the sustenance of the entire household depends. So, any decision concerning livestock management by the Abba Warra influence the livelihood security of the family and clan members who rely on him for good or bad. In households that are engaged both in livestock rearing and farming practices, the responsibility of the household head extends beyond livestock management and into overseeing farming activities. This entails high responsibility for the household head as he has to allocate his time between the various livelihood practices wisely. However, it has been observed in the field that the role of Karrayu pastoral women in handling the livestock within the household is becoming increasingly significant. The women participants from Giddara village in the focus group discussion elaborated on their increased role as livestock managers, once predominantly the sphere of the Karrayu men, in the following manner:

> In the past the men were the ones who totally take care of the livestock that we own in our family. However, many of the burdens concerning the managing of the livestock that stay near the homestead are on us. We collect forage, we treat the animals when they are sick, we milk them, we take them to the watering point. We literally do the job that only the men used to do. But these days, they are on the farm...some of them is in town...they simply stay in town and chew khat... [Women FGD, November, 2011].

In addition to the reasons given concerning the more important role of women within the pastoral Karrayu household, the increased diversification of risk management and livelihood practices that are outcomes of greater vulnerability due to social and climatic factors are also mentioned as factors that have distracted the attention of men from their livestock. Despite these new trends within Karrayu pastoral households in the study areas, it is the head of the household himself who oversees the everyday management of livestock, Furthermore, though the role of women in respect of livestock-related chores has increased, the head of the household still maintains the power to make decisions on matters related to livestock sale. The women focus group discussion participants from both study villages confirmed that:

> In our households it is the head (Abba Warra) who decides, together with the eldest son, which animals to take to the market and sell. The boys also help him with the difficult task of

taking the animals to the market place near Metehara town. After the sale it is also up to our husbands to decide how to allocate the money. Sometimes, they give us a certain amount to buy items for household consumption or clothes for the children (Women FGD, 2011).

In many cases, women in the Karrayu community get married at an earlier age. This age difference puts the man, with many years of experience in livestock rearing, in an advantaged position over his wife in making decisions at household level. However, through some culturally defined roles in household chores, the wife actively engages in activities such as milking cows, preparing food, and raising children that are very important in sustaining the entire household.

Within Karrayu pastoralists, the effective management of livestock, and thereby effective engagement in risk management and livelihood practices, involves almost every member of the household, but is highly gender and age-based. In table 4 below, I summarize the tasks undertaken by members of Karrayu pastoralists in Illala village, based on focus group discussions conducted with both men and women.

Table 4: Gendered and age-based roles within Karrayu pastoral households

Gender and Age	Responsible for	Tasks
Women above 50 and small boys and girls	Calves and new born sheep and goats	Tending the livestock within and around the homestead
Married women	Milking cows and unweaned calves	Maintaining and cleaning of calf pens; fetching water for calf consumption during dry season; allocation of milk to calves and people; forage collection
Unmarried women & young girls and boys	Cows with weaned calves	Taking livestock to watering and grazing sites not far from villages
Unmarried and strong young men and newly married women	Steers, bulls, young calves, and heifers and cows in their early gestation period	Defensive role and scouting by young men while the women support them by preparing food
Male adults, adolescents and boys	Camels	Long distance camel pastoralism; the young boys milk and keep inventory of the camels

Source: Compiled from focus group discussions at the two study sites (2011)

On top of the gendered and generational differences in the household, and the implications of this for decision-making on risk management and livelihood practices at household level, I will argue that the household best able to adapt is one in which members collaborate over making strategic decisions on managing their resources in order to better navigate the transformed and unpredictable natural environment of Metehara plain and beyond.

Under conditions of vulnerability emanating from social as well as ecological forces, pastoralists make major decisions concerning the type of livelihood and risk management practices that they should pursue and prioritize. Depending on the available resources, households engage in practicing ranging from livestock rearing to cultivation. Within livestock-rearing they also either specialize on particular species or keep mixes of different livestock species. In many cases Karrayu pastoral households follow the paths of their parents and take-up livestock rearing as their major livelihood practice. However, the frequent drought in upper Awash valley has reduced the number of livestock that forced households to consider cultivation in the course of time. Depending also on the worldview and level of education of the household head, families may take their own course and take up cultivation from the beginning.

During my research, I found out that most Karrayu pastoralists change the level of their pastoral engagement under the influence of broader socio-political and environmental conditions that they operate in. In line with this, an informant from Illala village mentioned that some forty-five years ago his parents used to keep around twenty milking cows and forty dairy goats. It was almost difficult to count the small ruminants as they multiplied fast because of the good rainfall and good pasture. As of now, however, his family owns only eight cattle and few goats. He associated the drastic dwindling of his livestock over the years to distribution for young siblings and own children upon marriage; necessary sales; massive livestock death from rinderpest, liver diseases and *furri* (a respiratory disease that attacks camels). He also further mentioned factors such as frequent drought episodes; land alienation by Metehara Sugar Factory and Awash National Park; conflict with Afar and Argoba; and expansion of Lake Beseka that made recovery of livestock almost impossible.

5.4 CONTEXTS INFLUENCING PASTORAL MOBILITY DECISIONS

Moving beyond the ability/agency of individuals to formulate and execute a decision, we must now look at the broader contextual factors that are not only taken into account when Karrayu pastoralists make migration decisions, but also those factors that limit or enable their ability to make such decisions in the face of climate stress. For example, the kinds of variables Karrayu pastoralists have to take into consideration depend on the resources they have available and the existing material conditions, including political and economic ones, which limit or enable decisions and actions. Clearly, social and economic position affects the decision-making process. Below I look at broader ecological, economic and political factors. At various times and for heterogeneously situated households these factors will have different impacts. In the study area, the Karrayu pastoral households try to evaluate environmental variables when scouting and selecting pastures for their livestock. Among the variables that they consider the major ones are topography, water availability, other communities on their way, and distance. Many of these variables also apply to everyday routine mobility practices.

Availability of Pasture
All Karrayu livestock keepers have a basic knowledge of grass and tree species and their relevance for their livestock during the various seasons of the year. However, preferences for specific grass or tree species depend on the type of livestock species that one keep. For example, during the informal discussions, some Karrayu pastoralists favoured chekorssa (flood plain pasture, Eleusine jaegeri) for cattle and goats while others favoured dhedecha (Acacia tortillis). For example, camel owners' and cattle owners' preferences for specific pasture and browsing differ. Boru Fentalle, a pastoralist from Illala village, further elaborated the issue in the following manner:

> All the families who are living in this village are the same because we are Karrayu. Except that some of us have many goats and sheep, others raise many cattle. As to my family, we are rich in the number of camels we own. So, we rarely stay in the homestead. Camels prefer the bushes and the salty soil beyond Haroole grazing site. The acacia trees in the village have fallen down due to many reasons so we have to search for a good browsing area for our camels. If I keep my camels in the village for long they do a lot of harm because they consume a lot. So, we do not simply move with all animals. We have to ask which forage is good for which animal. For the cattle, sometimes we collect the grass and cane top from the sugar plantation and feed them. You cannot do that with the camels [Illala village, 2010].

Karrayu pastoralists' knowledge of their ecological environment and the browsing and grazing resources vary considerably. Some of my informants go to the extent of which plants have a medicinal value to their livestock. In many cases, pastoral households asses the quality and quantity of pasture before they make decision on mobility.

Overall rangeland situation
During the time when groups of Karrayu young men (*salfaa*) go out to find out and asses pasture areas; their major consideration is not only the plant and grass species available in the rangeland but also the overall status of the pasture conditions. Infestations by non-palatable species prevent the choice of some pastures for grazing. For example, in Metehara plain some settlement and grazing sites are abandoned by the Karrayu herders because of the increased growth of Prosopis which goats will eat but no other species. They mentioned that Prosopis julifera is a sign of not properly managed and overgrazed pasture land. Other species can also cause problems. For instance, the infestation of nearby pasture and farmlands by Parthennium can force households to move or to forego

settlement. Parthennium can cover pasture in an extremely short period of time and it is not palatable to any of the species of livestock. The prevalence of this weed can also signify poor pasture and overgrazing. Overgrazed pasture can couple with events such as drought to produce severe conditions and Karrayu pastoralists always try to avoid overgrazed pastures. Of primary concern when pastoralists assess pasture conditions is the question whether or not there is significant pasture capacity. The gradual deterioration of pasture productivity definitely reduces critical areas customarily used as reserves for times of severe drought. During group discussion and informal interviews Karrayu pastoralists often concerned about the ever dwindling capacity of the rangeland to support their livestock. Such concerns were raised more frequently by those Karrayu households who keep relatively higher number of livestock than those who have less number of livestock. In many cases, those Karrayu households who are in possession of few cattle are much concerned about availability of water and grass than the overall condition of the grazing land. On the other hand, those Karrayu households who are in possession of more camels were concerned about the overall situation of the grazing land, browsing availability and salt licks as these are important factors that shape their risk management and livelihood practices.

General conditions of the livestock
The most important economic asset of the Karrayu pastoralists is the number of livestock they own. Due to the unpredictable natural environment, the livestock number fluctuates. These uncontrollable ecological factors influence the growth and decline of livestock numbers. Accordingly, pastoralists have to closely monitor the condition of their livestock. Such close monitoring of their livestock situations informs the Karrayu herders to take timely decision to move away from their current grazing site and relocate. This timely decision is very important as the biologically determined recovery phase of livestock after important drought or epidemic-induced losses have far reaching economic and social consequences (Bollig, 2006; Bollig and Göbel, 1997). In their assessments of the condition of their livestock, Karrayu pastoralists use some general conditions such as the body mass of their stock. The Karrayu pastoralists have to consider the effect of local environments on their livestock before they take any decision to move.

Availability and organization of labour
The availability and organisation of labour is an important factor in the risk management and livelihood practices of pastoralists' community. Hence, bottle-necks in availability of labour are important factors that influence the decisions of pastoralists in engaging in particular risk management and livelihood practices. The availability of household labour is also a decisive factor that the Karrayu pas-toralists consider before taking their animals to distant grazing sites. In particular the Karrayu pastoralists who own more livestock have to seriously consider the problem of labour shortage. In some cases, those households who are engaged in camel rearing try to solve the problem of labour shortage by making arrangements with livestock-poor households. A Karrayu pastoralist from Illala village who

owns a large flock of camels explained how he manages to deal with the problem of labour shortage in the following way:

> I own many camels and the problem that I have is the labour force. Out of my five children three are girls and the two boys are under fifteen. So, I arranged with two families from Qobo village to take care of some of my camels. They will take them to the distant grazing sites and handle them properly while they are using their milk. The milk is theirs. I do not have any problem with milk. I use the rest of the camels for that. Because of their need for cash they can also sell the milk. As long as they keep the camels in good condition and take them to suitable pastures, I do not expect the camels to be returned to me soon.

The problem of availability labour can be related to 'family development' – the cyclical change in the number and structure of members of the household to sustain a pastoral way of life. Here one may argue that the rich Karrayu pastoralists have better access to labour as they can easily form herding groups to counter the problem of labour shortage. It is also clear that vulnerability within a seemingly homogeneous community is determined by power relations that depend upon the possession of material resources. Though it is difficult to tell at this level that a patron-client relationship exists among the rich and poor pastoralists within the Karrayu community, it is clear that the livestock of rich Karrayu pastoralists can widen their access to resources. The practice of stock transfer is also worth mentioning because it indicates inter-territorial mobility and access to distant resources by utilizing the labour force of the poor pastoralists.

Another important arrangement of livestock entrustment among the Karrayu is *Dhebere*, a customary way of transferring milk livestock (Ayalew, 2001). This is mainly done on a mutual basis so that the household that receives the livestock benefits from the milk while taking care of the animals. It is also a social welfare mechanism through which the livestock-rich pastoralists give milk animals to poor households which take care of the animals and can retain the offspring. Through such an arrangement, pastoralists that have lost their core livestock due to drought and disease can revive their livelihood.

5.5 REORGANIZATION: CAMEL-BASED LIVELIHOOD PRACTICES

As discussed above, the transformation of the landscape in the Metehara plain has resulted in the disorganization of the social structure of the Karrayu pastoralists. Prior to the arrival of the commercial farms and protected area, the Karrayu were mainly practising a form of pastoralism centred on cattle herding. Cattle were central to their cultural fabric as well as fitting well into the vast semi-arid plain endowed with a variety of grasses. However, these changes in environment and the concomitant degradation of the pasture, combined with frequent droughts, have made cattle-based pastoralism almost impossible. This has pushed many pastoral households into poverty and made them dependent on food aid (see section 4.3.1).

Nevertheless, not all Karrayu households are sitting passively and waiting for their fate to be decided by others. As discussed in Chapter Two, actors actively employ their agency in order to survive even in very oppressive and harsh socio-environmental conditions. They utilize their power in order to work against the structural forces and achieve their own objective (Giddens, 1979). Accordingly, many households in the study site have reorganized themselves around camel-based pastoralism. In the following sections, I explore these camel-based livelihood practices as an adaptation to environmental change.

5.6 SURVIVING ON CAMELS: RISK MANAGEMENT AND LIVELIHOOD PRACTICES

Apart from the wealth of research on the evident ecological rationale of pastoral mobility, some scholars have questioned the highly determinsitic and naturalized perspective on mobility (McCabe, 1994). These critiques go beyond approaching pastoral mobility as response to a variety of ecological opportunities (Spencer, 1997). As Agrawal (1999:7) contends regarding pastoral mobility: "the normalization of mobility in the face of 'natural' risks may be misleading" because "mobility, far from being a simple, natural, and easily justified strategy to address fluctuations in biomass production" is rather "layered and tremendously complex in its origins and execution." McCabe (2004) states: "the way mobility plays out on the ground is far more messy and complicated than can be captured within this type of framework" as "all movements are balancing acts, influenced by ecological, social, and political factors" (237). Even though it could be one possible response to environmental stress, mobility is mainly influenced by social and economic factors that influence resource access. For instance, mobility can be used both as a means to deal with local scarcity of pasture and strategy to avoid some Socio-political sources of risk such as conflict. Gradually it is becoming clear, as McCabe (2004) points out, that: "Among ... contemporary pastoral peoples, access to markets, schools, medical facilities, and the extent to which they are integrated into regional and national economies exert a significant influence on the way they move and manage their livestock" (237). The crucial decision to move and to settle elsewhere is a multifarious and context-specific one and ecological concerns may be secondary. Moreover, the reason for moving away from one area and settling in another are often very different and possibly unrelated. Here, I focus on Karrayu camel herders and their mobility practices.

Camel-based risk management and livelihood practices among the Karrayu are responses to both environmental transformation and Socio-economic changes. Even though the cultural traditions of the Karrayu do not value camels over cattle, circumstances have forced them to raise camels as a means of adapting to the new ecological conditions. Camel-based pastoralism among the Karrayu is dependent on the knowledge and management of camel breeding. The extent that herd management affects household livestock wealth is determined by the underlying factors and conditions that shape household capacity for particular management practices.

5.6.1 Knowledge and herd management

Knowledge plays a decisive role in an individual's ability to actively engage in pastoralism as a livelihood strategy. The data collected from the pastoral village confirms this. Individual pastoral households in my research area have species-specific expertise. Many households, though, have their own specific preference for one species over another. Some Karrayu pastoralists are knowledgeable about sheep and goats, for example, but know little or nothing about camels. And some Karrayu households refer to themselves or each other as *'warraaa gallaa'* (villagers of the camel) rather than the more general term of livestock keeper. Although camels are more important economically than other species under current market conditions, some pastoralists are extremely cautious or even fearful of interacting with them as they have little or no familiarity with camels. Besides the preference of Karrayu pastoralists for specific species, an interesting observation during my field research was that the resources owned by Karrayu households also depend on the norms relating to those particular resources and hence influencing risk management and livelihood practices. As in the case of farming, which has a lesser status among some Karrayu households, camel herding is avoided by some groups in the Karrayu community for cultural and religious reasons. These people are called '*qallu*' and they are the spiritual leaders of the Karrayu traditional religion (*waqeefenna*). In the words of a Karrayu pastoralist from this group:

> We do not drink the milk and do not eat the meat of camels. For us we do not even see it as an animal. Its structure is strange and we as members of the *qallu*, if we have any contact with camels our *waqqaa* will be angry and we will be sick. So, we do not keep camels as livestock.

Despite the cultural priority given to cattle and the relatively recent reorganization around camel rearing, the Karrayu have a wealth of knowledge about their camels. They use various criteria to refer to their camels, such as age, colour and the role of the camel within the herd. The following table presents an example of how the Karrayu herders classify their camels in accordance with their age, sex and role within the flock.

Table 5: An example of camel categorization in the Karrayu community

Age	Sex	Characteristics/Purpose	Local name
4 years	M	Used for transportation of goods. They are kept at home away from the rest of the herd	Geejiiba
4–22	M	Bull for reproduction	Karabicha
4 years	F	A dam that has calved twice or more	Guuroo
12 years	M	An old bull that has been replaced by a young one	Karabicha dullacha
	F	Lactating camel with little milk	Bakkuu
	F	Lactating camel with a lot of milk	Mirgessa

Source: Based on group discussion in Illala village (2010)

As we can see in the table above, the camels that the Karrayu herders keep have different roles based on their age and sex. These different categories also have implications for management and labour arrangements. For example, the distinction between *Geejiiba* and *Guuroo* determines whether the focus is on the production of milk or on use for the market. Furthermore, as production of milk is crucial in the economic system, the Karrayu camel herders meticulously identify their camels based on their milk production. They need to know the level of milk production and categorize their livestock accordingly. This knowledge contributes to the production of more *Mirgessa* and makes the herders avoid choosing a bull that has been calved by a dam with little milk. It is believed that such a bull will produce only camels with little milk. This knowledge and method of categorization is an important part of camel management and influences the biological cycles of the herds.

Control over herd composition and size is an important element of herd management. One method is the culling of surplus animals such as adult males and unproductive females. The reduction of these categories of camels, in addition to generation of income, eases the pressure on the environment and keeps the right balance between animals, availability of human labour and browsing resources. According to my informants, dams and young female camels are the most numerous category within the herds. The second most populous category is calves between 0 and 3 years old. This composition of herds shows the deliberate and active role of the Karrayu in managing their herds. In addition to herd composition control, the efficient use of human labour and pasture is another management decision that needs to be addressed. Accordingly, the Karrayu camel herders organize themselves into camel herding management units (locally termed as *bulcha*). One camel management unit may be in charge of between eighty and five hundred camels, depending on the availability of labour and pasture. In many cases, those families who own camels come together to form one herd. This is a strategy to ease the problem of labour shortage. Those who have big families usually do not form herding groups with others, except in the case of security problems along the browsing route. My informants said that people are free to form herds with whomever they like, but generally speaking herd formation and the establishment of management units follow clan lines. People of the same clan or sub-clan usually form a herd together and the herd is named accordingly. However, despite the detailed knowledge of camel herding that they possess, many of my informants stressed that camel herding is a very arduous activity that demands high levels of labour and organization, as it is carried on mainly outside the traditional territories of the Karrayu.

Table 6: Camel herders' roles and responsibilities by age

Age	Roles and responsibilities
4–8	Children join the group to learn about camel herding. They also benefit from the availability of nutritious camel milk.
8–14	The herders in this category are the one who are responsible for milking the camels and carrying the milk container. They are responsible for counting the camels every evening and reporting to the elders.
14–25	Those in this category search for pastures ahead of the camels, build kraals, help the camels during mating and calving, and are responsible for the safety of the camels especially at night.
25–40	Responsible for the management of the herd and the labour force. All decisions are taken by this group, based on the information collected by the category above. Negotiations with any other group along the browsing route are also carried on by this group.

Source: Based on group discussion with camel herders (2010)

For this reason, the camel herders – mostly unmarried adults and young boys of different ages – stay away with their camels for most of the year. Culturally, women are not allowed to come near the camels, let alone herd them. Unlike their role in cattle and ruminant production, women are not even allowed to milk camels; camel milking is usually done by young boys between 8 and 14 years old. The following table shows the age categories and responsibilities of camel herders among the Karrayu.

Their way of life creates cohesion between the camel herders, and at the same time detachment from the core Karrayu community residing in the Metehara plain. In order to cement their group cohesion, camel herders drink from the same bowl to show their solidarity and as a way of promising not to desert each other in bad times. However, they also mentioned that sometimes camel herders compete with each other for pasture and water.

5.6.2 Access to browse: the routes of camel mobility

Until the 1980s, the camels of the Karrayu browsed in the vicinity of Metehara, such as Haroolee plain and places not far from Metehara such as Chercher. The Karrayu then started to move with their camels towards the west and southwest of the Metehara plain and into the highlands. The dry season areas include the highland districts to the west and southwest of the Fentalle district (Karrayu land). It is widely known that mobility among the Karrayu pastoralists is based on practices that are routinely incorporated into the social fabric. However, the changing social and environmental conditions in Fentalle *Woreda* have forced the Karrayu to re-evaluate their alternatives to mobility as an adaptation strategy. The Karrayu pastoralists keep various species of livestock. Among these species camels have played an important role in recent times as they serve the dual purpose of utilizing distant pastures as well as providing milk and cash when they are sold. The camels that the Karrayu pastoralists raise browse on nearby grazing lands such as Haroole plain. However, the browsing resources in the nearby villages have declined due to various factors such as the expansion of Metehara Sugar Plantation, Abadir cotton farm, Nura Era plantation and the expansion of Lake Beseka. All these factors have obliged the Karrayu to search for alternative grazing sites for their camels, moving out of the nearby areas to find suitable browsing. The decision to seek 'other' grazing areas is significantly influenced by the factors that I have discussed in the previous sections. My observation of camel-herding pastoralists shows at least three types of mobility that are being practised by these pastoralists.

Type I: Short-distance movement
Herders move to the customary grazing sites within their territory as a kind of default. All the four households that I studied practise the customary mobility patterns of their villages in Illala, travelling to the nearby Haroole plain and to the adjacent Boset *Woreda*, an journey that may take them up to three days. This kind of mobility, practised not far from the Karrayu territory, covers a maximum distance of forty kilometres. Since almost all of these customary sites fall within the Karrayu territory, the Karrayu herders consider fewer factors when they move to these grazing sites. Instead they concentrate more narrowly on ecological conditions and their livestock needs. Only when the moves are of greater distance do other limiting factors play stronger roles.

Figure 15: New forms of mobility by camel herding Karrayu pastoralists

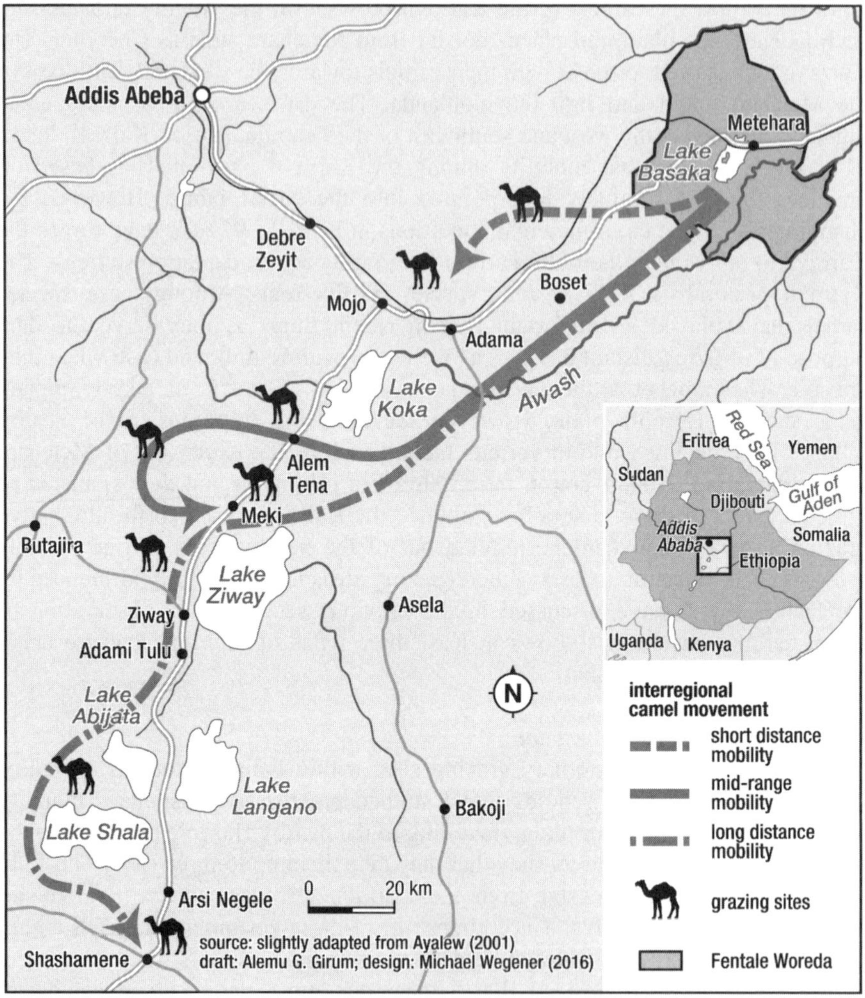

Type II: Mid-range camel mobility
The second type of mobility practised by camel herders is travelling out of their own traditional territory but remaining within what they call the territory of 'friends'. They refer to such areas as Melak Jilo, Sodere, Alem Tena, Koka and Walanchiti areas. They consider browsing sites in the vicinity of these territories as friendly because they are mainly inhabited by Oromo-speaking groups with whom they can easily communicate if their camels create damage to other people's property. According to these pastoralists, they also try to benefit from the abundance of salty spring water in the vicinity of Sodere which their camels really like. The estimated distance that they cover is between sixty and eighty kilometres

and the journey may take one to two months. In addition to the longer distance that the case study pastoralists travel, inspecting pastures and guaranteeing access to resources like water and good quality pasture are critical. Moreover, as with any distant grazing sites, this requires additional labour, particularly so as my case study households keep livestock of various species.

Type III: Long-distance mobility
According to my case study households, long-distance strategies of camel mobility require more contemplation than the other two types. This is the case because they have to consider the farming communities that practise a totally different kind of resource utilization. The destination areas mentioned by the households include Adami Tulu, Meki, Lake Ziway area, Bulbula, Lanagno, Abiyata, Shala, Arsi Negelle and Shashemene. Along all these routes, there is a high risk of conflict with the cultivators as their camels may intrude on the farm lands and create damage. With such long-distance routes that criss-cross vast areas dominated by agriculturalists, they try to limit the camel herds to a smaller size.

Table 7: Summary of mobility types as practised by the camel-herders

Mobility	Distances	Destination sites	Constraints
Short	Up to 40 km	Haroole plain, Mount Fentalle area, Awash National Park (within the Fentalle *Woreda*) and near Boset *Woreda*	Lack of appropriate grazing pasture
Mid-range	60–80 km	Sodere, Koka and Walanchiti (friendly territories)	The problem of additional labour
Long	100–200 km	Lake Ziway area, Bulbula, Lanagno, Abiyata, Shala, Arsi Negelle and Shahemene	The problem of additional labour, the risk of conflict with farming communities along the route

Source: Based on case study pastoralists in Illala village

Furthermore, as they cover distances of between 100–200 kilometres, all the case study households said that they start from the Fentalle territory around September and browse their camels on the way to Shashemene, embarking on the return journey around April in order to avoid the beginning of the agricultural season in these areasAll in all, such a long journey lasts six to eight months. When they arrive back in their area, they keep their camels in the plains of Haroole till the next journey starts in September again. This strategy, though labour-intensive and demanding for the household members involved, especially because of the social distance, permits herders to cope with the patchy distribution of browsing resources and respond quickly to adverse conditions. Overall the pastoralists are significantly more mobile than in the past. During the first and second phases of my field research (February–March 2009 and July–September 2010), I made a study of mobility in the case study households which showed that in a normal year these pastoralists are significantly less mobile than they were during the first field research period (i.e. February-March 2009). Nevertheless, the pastoralists reported that over time the frequency of mobility and average distance is on the rise. However, increasingly herders have had to develop new strategies for increasing their access to pastoral resource not only to fulfil the requirements of larger herds and increased production but also to circumvent what herders see as an increasingly risky environment.

5.7 SUMMARY

In this chapter I have shown that the environmental transformations and the associated disruption in the social organization of the Karrayu community. The changes in environment-pastoralists interactions have implications for the risk managemnt and livelihood practices pursued by the Karrayyu. Throughout the chapter it has been shown that the changes in customary mechnaisms of managing natural resources resulted in changes in mobility practices of the Karrayu and thereby seriously hampered the primarily cattle-based pastoral livelihood of the Karrayu. The Karrayu households employ flexible strategies, especially mobility of different kinds, in response to the changing socio-cultural and environmental conditions. I take a critical stance in respect of the common understanding of vulnerability reduction strategies used by pastoralists, which tends to see them as a simple response to the unpredictable environment. By zooming in at the individual herders level, this chapter has revealed that pastoral strategies for managing risk and securing livelihood are based on complex considerations involving various socio-cultural and political factors, in addition to the ecological factors. The reorganization around camel-based livelihood practice and the continuity of pastoral way of life amoung the Karrayu are also discussed in detail. Moreover, I have explained some of the factors that affect the way pastoral actors use mobility as a risk management and livelihood strategy. Various constraints limit the timing, distance, and end location of their mobility. In the context of climate change, lack of access to grazing sites and therefore mobility as a vulnerability reduction strategy, can harshly influence the ability of the pastoralists to exercise their risk management and livelihood practices optimally. In the following chapter, I will show how the Karrayu pastoralists use their agency within the limited spaces of adaptation by shifting to cultivation as a risk management and livelihood practice.

6. LIVING THE TRANSFORMATION: THE MOVE TOWARDS AGRO-PASTORALISM

"We always want to increase the number of animals that we raise. Have you noticed the market price for goats and sheep in Metehara these days? We will also use the irrigation to farm but we will not replace our livestock with it. We say that a rat that has two holes will never die!" [Group of pastoralists, Illala village]

6.1 INTRODUCTION

As has been discussed in Chapter Two, pastoralists actively employ their agency in order to deal with the structural sources of vulnerability. The 'duality' between structural forces of vulnerability and pastoral agency leads to the formation of emerging practices through active engagement by the actors. Consequently, rather than viewing actors as unintentionally responding to structural shifts in power relations emanating from above, this research takes seriously the notion that local resource users are agents in their own right. However, this "requires a concept of human agency that is neither determined by social structure nor entirely voluntaristic" (Gezon, 2006:15). Actors, in this framework, are seen as socially positioned with systematic and patterned affinities and dispositions (Rocheleau, 1995), while at the same time, seeking their own benefit, constantly engaged in negotiation and "the work of social change" (Bailey, 2001). Actors are also situated within differential, yet dynamic, power matrices, where identities such as gender and ethnicity and other collectivities frame strategies and shape social interaction. Giddens also takes the concept of agency and defines it as "the capability of an actor to act otherwise". In this chapter I explore how the Karrayu pastoral groups have managed to actively balance their traditional livestock-based mechanisms of dealing with vulnerability with cultivation in the process of interacting with their environment and the Ethiopian state.

6.2 CULTIVATION AS RISK MANAGEMENT AND LIVELIHOOD PRACTICE

Among the Karrayu pastoral groups, the new forms of risk management and livelihood practices have evolved over the years, primarily due to the changing socio-cultural sources of vulnerability, such as population pressure and loss of grazing land, rather than as a direct response to climate change. The local groups do not make a distinction between their traditional and transformed ways of dealing with vulnerability. However, one can observe that in some cases traditional livestock-based mechanisms have changed to contemporary forms, parallel to changes in culture and everyday life. These emergent forms of handling vulnerability can serve purposes beyond merely buffering individuals against risk, for they also give them orientation regarding the future of their livelihood trajectories. This shows us how the locals try to actively appropriate, transform and make use of resources and power structures that are external to the community. These external resources are important, as the locals' mechanisms of dealing with climate stress have been eroded over the years. In this regard, by expanding the conception of power to active agency at various levels, while appreciating the social nature of agency, political ecologists take into consideration the diverse range of factors that influence the knowledge, decisions, and actions of the locals in their attempts to access and use resources and thereby deal with vulnerability arising from climate stresses. Clearly, the conceptualization of political ecology stated here gives primacy to the real social world of interacting agents. In Chapter Five, I have given emphasis to the routine, iterational pastoral agency incorporated into livestock-based activities that the locals employ in order to handle vulnerability. However, with the increasingly constrained spaces of adaptation, the pastoral way of life has changed its form in the process of assessing the resource base and institutions that have supported local pastoral practices in the past. As has been shown in Chapter Two, the practical-evaluative aspect of agency can be seen in situations where the emerging dilemmas and ambiguities of presently evolving conditions require actors to make practical and normative choices among alternative possible trajectories of action. Actors' practices result in consequences, intended and unintended, that reproduce, perhaps with modifications, structuring principles. The actor's reflection on consequences feeds into the consciousness of the actor. The most deeply layered or deposited practices, those actions that are reproduced over and over again, become sedimented practices of local adaptation.

Unless the members of the household own only small stock and therefore concentrate on cultivation, or they own enough livestock to become exclusive pastoralists, each Karrayu household try to balance the two activities. Those with moderate herds need to decide each year whether or not expending labour on cultivation is in the best interests of the household. When I asked people from cultivating households which was stronger in their household, cultivation or herding, most men and women answered 'cultivation', but some answered 'both'. People from households with larger herds, or a strong predisposition toward pastoralism, answered herding: their livestock carried them through the year, and cultivation supplied them with only a small part of their food.

In her study of how households move between a sedentary cultivation way of life and that of exclusive pastoralism, Mace (1993) only considered the amount of grain that a household can store and did not take into account in-between situations that mediate the transition process. However, in the reality of the arid and semi-arid environments, harvests are temporary, the stored grain more often than not lasting only a few months. One cannot base a decision to become sedentary on the amount of grain stored; rather the decision to become mobile is

6.3 SIGNIFICANT ASPECTS INFLUENCING DECISIONS TO CULTIVATE

There are a number of historical as well as contemporary factors that have contributed to the shift towards agro-pastoralism among the Karrayu. In broader terms, there are three important and interrelated forces: internal social differentiation, population pressure, and policy issues and the push towards irrigated farming.

6.3.1 Disruption of social relations and increase in farming

As stated in the previous chapter, for several decades the Karrayu were exclusive pastoralists, deriving their sustenance from rearing different livestock species. Nevertheless, within the last six to seven decades, due to a combination of broader Socio-political and ecological factors, the role of the pastoral way of life based on livestock production as a sole source of livelihood has gradually declined. The main reasons for this are the shrinkage of the resource base due to external interventions and consequent environmental degradation, and the recurrent drought.

Development interventions like those of the commercial farms were not favourable to pastoralists in the valley, since this resulted in the alienation of the dry season grazing lands of the Karrayu pastoralists (Ayalew, 2001; Müller-Mahn, et al., 2010). The Karrayu case is further aggravated by the Ittu migration and the establishment of the Awash National Park and the expansion of Lake Beseka. These developments have led to increased vulnerability of the pastoral way of life by reduced access to resources that are crucial in supporting mobile strategies. The loss of pasture and the recurrent drought have resulted in a sharp decline in the size of the livestock population[5] and impoverishment of the Karrayu.

All these processes of vulnerability have forced the once full-time pastoralists to adapt themselves flexibly to agro-pastoralism in which the herdsmen combine livestock raising with farming practices as an adaptive measure. Thus, a supplementary non-pastoral activity, namely farming, began to expand despite unsuitable climatic conditions. Farming is also dictated by the need to supply the increasing population with food, which the livestock sector alone cannot achieve.

---

5 The mean number of livestock per household plummeted from 59.4 TLU in 1977 to 14.5 in 1997 (Tibebe, 1997, cited in Ayalew, 2001).

In the context of the Karrayu, cultivation is practised in two different ways, namely, rain-fed and irrigated (Ayalew, 2001). Both mechanisms of crop production in Karrayu land have been used for the past three decades. The Karrayu themselves have not developed an agricultural tradition. Agriculture was introduced to the Karrayu areas in the wake of the migration of the agro-pastoralist Ittu. According to the Karrayu elders, it was introduced by the Ittu migrants, especially by those agro-pastoralists who came to Fentalle district for the first time during the 1984/85 drought due to the crop failure in West Hararghe.

Photo 4: The practice of farming in Karrayu community

Source: Author

Before 1984/85, the Ittu migrants were only pastoralists. Most of the Ittu agro-pastoralists who came to the area in 1984 settled near the sugar cane farms on the banks of the Awash River and started small-scale irrigated farms, later expanding to other villages in the Karrayu territory. According to the focus group discussion participants in Giddarra village, irrigated farming began first, using small water pumps. Then, rain-fed farming started on the foot-hills of Mount Fentalle. According to the focus group discussion participants, rain-fed farming was practised for the first time in Kobbo, followed by Dhaga Heduu and Laga Banti (Ayalew, 2001). The Karrayu pastoralists in Giddara village with whom I held the focus group discussion emphasized the decline in livestock numbers and the high price for grains as the major reasons for beginning rain-fed farming.

The key informants further added that the agro-pastoralist Ittu who migrated to the Karrayu area during this time convinced the Karrayu that their country is conducive to rain-fed farming. The Ittu asked the Karrayu why they starve while they have this opportunity. So the scarcity of grazing land to support the Karrayu livestock, which held them back from fully applying the traditional pastoral mechanisms of dealing with vulnerability, and the increasing awareness of farming that they got from the migrant Ittu, encouraged them to start crop production. The informants explained the influence of the Ittu on their current situation as follows:

> It is the Ittu who started cultivation on our land. We slowly followed them. In the beginning we used hoes but later we began to use oxen as draught animals. It is also the Ittu who used to train our oxen for farming and make the yoke and other farm implements for us.
>
> We have now started to farm because we see our Ittu friends build a nice house with corrugated iron…when you have a farm you need to be next to it…some of them have TV and also own small restaurants.
>
> In the past, we used to keep many animals because the grasses and water were available. The men used to take care of their hair with style…sometimes we go to the borders to Afar and see who is more brave and daring to come near to each other. Now, we don't have time. The farming makes us busy. We wear shorts unlike the past and we cut our hair short because we are busy.

This shows us how individual actors are actively engaged in the negotiation and production of social change while at the same time pursuing their own benefits. This has far reaching implication for their risk management and livelihood security. Maize cultivated by rain-fed and irrigated agriculture is the major crop at my research sites (Illala and Giddarra), as in other Karrayu *Kebeles* such as Galcha, Gara Dima, Dhaga Hedu and Xuxxuxi. For many Karrayu pastoral families, it serves to fill the consumption gap that the pastoralists cannot satisfy using livestock products only. My informants are also in favour of planting maize because in times of good harvest it saves them from selling livestock to buy grains. Moreover, maize provides residues for livestock after harvest, even when they face crop failure. Accordingly, maize has become the staple diet of the Karrayu since the change of their traditional consumption pattern, which was based on livestock output, due to the various interlinked factors associated with external interventions.

The crop cultivation practice that the Karrayu pastoralists have adopted has several purposes. In addition to satisfying household consumption needs, it also helps households to maintain or increase their herd size by reducing the number of animals that need to be sold in order to buy grain. My informant from Giddarra Village, Hawas Bulga, said the following:

> In our eyes cultivation was highly despised. We preferred to sell animals and buy grain rather than farming. But, when we saw the benefit of farming from our Ittu brothers, we said to ourselves…ahh…farming is not bad and there is nothing wrong with farming…but many of us could not accept it easily…I remember in the beginning no Karrayu was willing to be identified as a farmer…but later we discovered that maize not only fills our belly but also fodder for our livestock…it really helps us not to sell our animals easily…

However, in spite of improvements in understanding the importance of farming, lack of proper knowledge about the practice of farming is one important obstacle that the Karrayu mention as a problem. In line with this, some authors, such as Muderis (1998), associate the low level of crop production in Karrayu land with lack of experience in farming, and absence of assistance in the form of access to fertilizers, seeds, extension advice and farm tools. Where they are engaged in cultivation activities, many of the Karrayu pastoralists prefer to do it through one or other form of arrangement with the experienced Ittu farmers. The size of the irrigated farms owned by Karrayu pastoral families is variable, from half to one hectare. Women do not own farms alone, but only together with their husbands. Each household head cultivates his farmland with family members. Some household heads form a work party in which they collaborate with each other to cultivate their land in turn, through the traditional collaborative institution called *jigge*[6]. In almost all the villages that I observed, the main products of these small irrigated farms are maize and marketable vegetables, such as tomatoes and onions. In Giddarra, in addition to the above, *khat*[7] is also grown. The diversification of crop types increased with the arrival of the irrigation scheme. In the past, very few households planted crops other than maize, due to inadequate water discharge from the plantation. If there is no problem of water supply, there may be three harvests in a year.

In Fentalle *Woreda* in general, and well-watered villages in particular, a new concept came to the Karrayu community with increased farming activity: private land, which was not a part of the communal land ownership philosophy of the Karrayu and their customary practice. As a result, I observed that large plots of land are currently being enclosed in the name of farmland. Here, I would like to argue in line with Giddens who argues that the change in meaning with regard to resources and the resultant steps taken by active agents have influenced the social structure of the Karrayu pastoralists over the years. In line with Gidddens' notion of 'duality', this shows us how human agency produces social structures which in turn are the medium for their own production and reproduction, again through human agency or practice (Giddens, 1976:121). For instance, the small-scale irrigated farms that have been set up on communal lands mean that land has been enclosed privately for year-round farming activities. This has clearly resulted in increased private ownership of former communal land that could have been used as pasture land. However, as those who are practising farming are also indirectly supporting the state's agenda of sedentarization, pastoral households are encouraged to enclose land for private purposes. For instance, during my field work in 2011, I observed around forty Ittu farmers who had been brought from West Hararghe and settled in a Karayyu village called FateLedi so that they could help to disseminate farming among the Karrayu.[8] These social changes also have

---

6 Jigge is reciprocal cooperation by farmers for conducting the farming activities of each member of the work party by turn.
7 *Khat* is a stimulant when it is chewed, and is a major cash crop among the Ittu farmers in Hararghe. It is believed that this plant was introduced to Fentalle *Woreda* by the Ittu.
8 The Ittu are considered by the local officials as instrumental in pushing the agenda of introducing more farming practices across the pastoral villages due to their close inter-marriage ties with the Karrayu.

6.3.2 Population pressure, settlement and farming practices

In the previous chapter, I have mentioned the changing norms and rules concerning resource utilization among the Karrayu and the new meanings attached to resources. In the focus group discussion with those who practise farming beside a pastoral way of life, it came out that in addition to livestock, a household head needs to have land for cultivation to make money and become successful. They understand being modern as being a 'model' farmer, thus gaining more profit, having a house with a corrugated iron roof, buying a small three-wheel motor car to provide transport services in town, and many more ambitions. However, not everyone in the Karrayu community benefits from the sedentarization process. For instance, the shift from communal land ownership to private ownership of plots has reduced the role of women in controlling resources. With less livestock and splitting of families, sedentary females may lose property rights, as land will continue to be handed down in the father's clan. In sedentary villages, a decrease of animals may also mean that women have less access and control over lactating cows, and are less able to sell animal produce.

In the words of a participant in the focus group discussion in Giddarra village, "We see all these successes achieved by the Ittu right in front of our eyes and we want to become like them by planting cash crops and selling them when the price is good without affecting our livestock number". The association of farming and a sedentary life with modern household appliances such as TV and refrigerator is also prevalent among the Karrayu, particularly among the younger generation. During the focus group discussions in Giddara village, statements like "We always aspire to work hard on our farms and get more money…buy Isuzu…become a model farmer" were very common among Karrayu pastoralists. In this regard, the pastoral actors are using their projective agency beside their practical-evaluative agency. Hence, the separation of the three elements of agency is very difficult in actual everyday practice of Karrayu pastoralists.

Narratives on matters of property and access among pastoralists are at two opposite poles. On the one hand is the argument for a general and undisputed recognition of common property as the institutional foundation of mobile pastoral communities, and on the other an emphasis on how such institutions have been taken over by state-enacted or development-induced private claims. What follows here contests the dichotomy of common property theory and the linear pathway of privatization narratives. In line with the political ecology approach, the meanings that the various Karrayu actors give to resources have changed over time as they interact with other groups such as the migrant Ittu population. This change in meaning of resources is crucial not only because meanings shape actions but also because certain meanings become manifested in material ways that shape the possibilities for interventions by others. I explore the kinds of rights Karrayu pastoral households claim over resources, and new pathways for exchanging resources in order to deal with the socio-cultural and climate-stress related sources of vulnerability that are discussed in Chapter Four.

The political ecology approach sees the problem that the Karrayu pastoralists are currently experiencing as emanating from extreme climatic events that reduce the livestock base as well as the condition of the pasture and their integration into the wider socio-political system, and the subsequent policy interventions by successive Ethiopian governments which resulted in the loss of traditional Karrayu pastoral autonomy. However, the resource access problem faced by the Karrayu is not only attributed to the direct loss of grazing land due to the establishment of the commercial irrigated farms and the Awash national park, but also to population pressure.

As discussed in Chapter Four, the problem of decline in pastureland was exacerbated by the demographic pressure that came about with the continuous encroachment and eventual settlement of the Ittu[9] on Karrayu land. However, despite the various sources of vulnerability to which the Karrayu pastoralists are exposed, they are not passive agents that sit and watch as various disasters happen to them. Rather, they actively utilize their 'pastoral agency' to deal with the sources of vulnerability. The different Karrayu pastoral families try to actively engage with various structural problems in order to come up with new mechanisms for handling them. It is not only the climatic sources of vulnerability but also the socio-cultural sources of vulnerability that have resulted in new forms of scarce resource utilization in the Karrayu community. This shows us that within the broader structural constraints that the Karrayu actors are faced with, they apply their agency, and try to combine the available resources and options so as to modify or get along with the system. One actor alone, of course, cannot produce or reproduce a social system. Indeed, a social structure or system implies interaction between actors. Actors use their own stocks of knowledge and capabilities, and are constrained by rules, when interacting, thus recreating the structure that those rules and resources constitute (Giddens 1979:71). The actors reflect upon the intended and unintended consequences of their practices and interactions, adjusting their reactions and thus modifying the structure.

---

9   According to Fentalle *Woreda* Agriculture and Rural Development Office, around seven villages out of the eighteen in the *Woreda* are dominated by the Ittu who have migrated to the area over the past several years. In five of the seven villages where the population of the Ittu is high, mixed farming and agro-pastoralism are the major forms of subsistence.

6.3.2 Policy factors: irrigation and land certification

As discussed in the previous chapters, the Ethiopian government is biased towards settling the pastoralists. The emphasis of the government policy is on a crop-based farming system that does not take into consideration the pastoral ecosystem and gives minimal attention to the livestock sector. While there seems to be a development policy to address and promote the pastoral production system, the establishment of farmers' training centres in pastoral areas shows that training and extension services target crop-based livelihoods at the expense of extensive livestock production. The introduction of the Fentalle irrigation scheme is a reflection of the government's preference for the crop-based sedentarized livelihood system, with the unintended consequence of tenure insecurity, as it has attracted large numbers of landless people into the area. Sedentarization is a social engineering scheme; it not only transforms the mode and means of production but also drives social and cultural changes (Scott, 1998).Those people who come to the area looking for plots are attracted by the irrigation scheme. They do not have any interest in the pastoral way of life and do not abide by the customary rules. Rather they are in dire need of a plot of land that can be cultivated. This has resulted in the appropriation of communal land for farming purposes. The privatization of communal land for farming purposes has generated a sense of tenure insecurity and led to conflicts. As a result, the Karrayu households who were scared of losing the rest of their land started to engage in farming as a livelihood strategy and thereby laid claim to the land. The focus group discussion participants in Giddarra village mentioned that the government had caused the proliferation of farming by encouraging the Ittu farmers through provision of training and agricultural inputs. So they decided to take up farming as a way to counterbalance the ever increasing enclosure of land by the Ittu and as a way not to be left out of the government's project. However, in spite of the increase of farming activities in the Karrayu community, pastoralism is still the main form of livelihood for many Karrayu households. They undertake both farming and pastoralism simultaneously. The Karrayu have been practising irrigated and rain-fed agriculture over the past twenty to twenty-five years. The most important reason behind their move to crop farming is the increased loss of their livestock due to drought and the shrinkage and fragmentation of the rangeland. In response to these problems, many Karrayu households in the study area have engaged in farming activities, both as a form of risk management and as a livelihood strategy. In earlier times, rain-fed agriculture was practised by different households among the Karrayu as a coping mechanism. However, such a practice was unreliable due to the erratic nature of rainfall in the semi-arid environment of Fentalle *Woreda*. On top of the drought and ever shrinking pasture land which have affected livestock husbandry and led the pastoralists to take up farming activities, the arrival and settling of the Ittu who migrated from West Hararghe has resulted in competition for the scarce resources. This has pushed the Karrayu to turn to the cultivation of maize. Furthermore, the fear of losing their remaining land if it is not cultivated played a role in the decision of Karrayu households to take up farming as an economic adaptation and diversification strategy.

In recent times, the Karrayu pastoralists in the villages of Giddarra, Garadimma, and Xuxxuxi have started to practise farming through irrigation. This is due to the introduction of a diversion canal that takes the Awash River to the Karrayu villages. This intervention has helped the Karrayu pastoralists to intensify their involvement in farming activities. One of my informants in Giddarra village described the mixed feelings of excitement and worry:

> We were really surprised to see the water crossing our village like that. In the beginning we thought that they wanted to give the remaining part of our land to the sugar factory as they did in the past. But, now it is ours. We started to produce many crops such as maize, onion and cabbage.

Despite the contested and different views of the practice of farming among Karrayu pastoralists, I argue that the process of taking up farming as a risk management and livelihood security strategy within the Karrayu community is an outcome of the general shift in traditional vulnerability reduction tools, which were exclusively dependent on livestock, due to social as well as climate-related sources of vulnerability. Irrigated farming has been expanding in many villages of the Karrayu community. In general terms, beside the increase in abundance of water in the vicinity, the shrinkage of pastureland to support extensive livestock herding was mentioned as a pressing factor behind the decisions by the Karrayu pastoralists in the study areas to take up farming. Furthermore, the increased contact with other farming communities, the transfer of skills, and the role of new organizing practices that facilitated farming also played a role in the increase of farming as an emerging way of handling vulnerability.

With the rise of farming as a livelihood strategy came new forms of organizing labour and other resources. With the increased conversion of communal pasture land to individually owned plots, the search for cultivatable land has increased. The Karrayu pastoralists have started to use their farm enclosures for the cultivation of maize and other cash crops, such as onions. They also lease out portions of enclosed land to other groups from the nearby towns of Addis Ketema and Metehara, who want to hire land for cultivation on the basis of sharecropping arrangements with the Karrayu who claim the land on an individual basis. In the study sites the sharecropping arrangements take four different forms and are completely different from the previous Karrayu pastoral mode of land arrangement. Rather they resemble arrangements that are common in places where farming is a dominant activity.

Photo 5: The new irrigation scheme has increased water availability

Source: Author

Increasing competition to enclose arable land has led to the shrinkage of the previously communally managed resources. This is further catalysed by the recent introduction of the new irrigation scheme. During my field work in 2011, in line with the on-going privatization of land by the locals, I also observed the local administrators distributing land in pastoral villages as a preparation for the in-coming irrigation facility to the villages. One of my informants, a development agent (DA) in Illala village, explained that:

> We are now rationing land on the basis of household heads. Each household head will receive a total of 0.75 ha of land for farming purposes. We also distributed a total of 0.5 ha of land to men who are above 18 years old but not married.

However, the Karrayu household members whom I interviewed are unhappy with the distribution process as it does not consider several socio-economic factors. One important element that they mentioned is the number of households that the male head manages in a polygamous community. They argue that a household head gets a parcel of land based on only one wife. In addition, the land distribution scheme is criticized as gender-blind since it does not consider women as having any rights over land, except that they may utilize the land of their husbands. In this regard, the government's push for sedentarization by allocating private land holdings confirms how institutions differentially empower and constrain actors' choices by supporting and privileging particular norms, rules, and structures, and are "clearly not simply mechanisms for efficiently allocating resources" but are also techniques of discipline and control (Agrawal, 2005). This confirms the argument that access to critical resources such as land is shaped by pre-existing social structures that try to perpetuate the system on behalf of the powerful. The herders who possess a large number of livestock also contend that the distribution of 0.75 ha of land to each household is unrealistic. Haji Nura Xinno stated:

> We have many cows, goats and sheep in the village. They are telling us to confine ourselves and the livestock in 0.75 ha of land. How is that possible? We are happy that the water is here

but what they are telling us is to reduce the number of animals that we own and start farming. We farm, yes, but we will also keep our animals. There is no guarantee for us that we will prosper through farming. We say that a rat that has two holes will never die! (Field note, May 2011, Illala village)

The Karrayu pastoralists fear that they will lose their pasture land to private farming enclosures. The arrival of the new irrigation scheme has accelerated the speed at which land is privatized. However, such responses by the Karrayu pastoralists do not necessarily lead to disparity and accumulation of wealth in the hands of the few. Those who have managed to produce more than what they need for their own consumption generally distribute the surplus to related families in distant villages where the irrigation water has not yet arrived. This is condemned by the local officials who want the Karrayu pastoralists to have not only private land but also a private life based on the concept of capital accumulation.

Photo 6: Fencing off communal grazing areas for farming purposes

Source: Author

Today, the Karrayu pastoralists who have a sizable number of livestock are concerned about the privatization of land for farming purposes and the lack of attention that the livestock sector has received from the government. Accordingly, they have also enclosed a sizable portion of communal grazing land, keeping out animals belonging to other herders and enabling them to graze their own stock. In other cases, individuals have fenced off extensive areas as farmland, although they are not always able to exploit their enclosures effectively due to their large size. The recent land distribution by the local administrators is an attempt to redress this problem and secure equal access to land by every family. This led them to give land of a fixed size to each pastoral family head. Although herders have access to the newly enclosed farmlands as there are no proper fences around them, the land still belongs to the individual who puts up fences and plants maize on it. Other community members in the villages where private farming has already started forbid attempts by others to privatize the area by saying that the land belongs to other occupants.

However, despite the tendency of Karrayu households to enclose the communal grazing areas for private purposes, grazing is allowed for all animals as long as the ownership rights of the man who enclosed the land are recognized. The first man who manages to enclose the land is entitled to use it for the grazing of his own stock or for cultivation. The proliferation of fencing represents the vast competition for land and the feeling of insecurity that one day they may lose the land that came about largely due to encroachment by outsiders.

For the time being Karrayu households are benefiting from a *de facto* private access resource regime. Most households have small plots of farm land and favourable ecological conditions due to the availability of water in the normally dry areas of the district. The Karrayu households with whom I conducted focus group discussions argued that there are few disputes concerning resource use. Other research, however, has reported increased disputes as a result of trespassing (Ayalew, 2001). Yet, in the current situation it seems there is wide-spread recognition of possession rights to farmland and grazing enclosures. On the other hand, there is a tendency to revive customary institutions that try to maintain and manage communal holding by introducing a communal enclosure for livestock. However, as the Karrayu pastoralists are in the midst of experimenting with the contemporary ways of handling vulnerability, they still have to deal with institutional ambiguities. Experimenting with new practices takes time to sediment and become part of the traditional mechanism of managing risk and securing livelihood, which in turn may influence pastoralists' vulnerability to environmental and climate-related stresses.

People began to enclose areas of land both in a bid to protect their individual plots from trampling and as a response to the new irrigation scheme. The enclosure helps them to grow suitable pasture for their own livestock. There are Karrayu households who have enclosed more than two hectares of land once considered to be a communal grazing area as private pasture for their domestic herd.

In the dry land areas of the district, more households have access to land exceeding one hectare, while at the well-watered villages many have holdings of less than one hectare. This is a clear indication that land scarcity is more severe among the farmers in the agro-pastoral villages due to the increased population pressure there. Basically, the Karrayu are well aware that the land scarcity problem in their locality has happened as a dynamic process explained by a number of interrelated factors. They complain very much about the inadequacy of land as one of the hindrances to expanding their crop production and raising livestock in large numbers. The perceived changes and dynamics were expressed by *abba gada* Kawoo, an elderly informant, at Giddara in this manner:

> The highly overcrowded land you see now was quite wild fifty years before. The natural vegetation cover was very dense, the settlements were very scattered, and hence most parts of the land remained untouched. Today, you see every corner being put under cultivation. In the past, the milk that a single cow used to yield could be sufficient not only for the owner but also for the neighbours, because there was no problem with grazing lands and no restrictions in the movements of livestock. By contrast, today even the so-called rich households possess a

pair of oxen, while in the past I knew households who had more than five pairs of oxen. This does not mean that all were used for draft power. So, the change I perceived in relation to land in our locality is quite excessive and considerable. Resulting from this, the social relations among the people have also undergone much change. [Field note, 2010]

Rain-fed crop production by the Karrayu began in response to the pastoral livelihood crisis, and is also practised on lands enclosed by individual pastoralists. As noted by Ayalew (2001), my informant Nura Qumbi confirmed that in Illala and other *Kebeles* such as Xuxxuxi, Dhebiti and Haro Qersa, communal land is currently being enclosed for use as private grazing and cultivation, and to protect the land and resources from being further alienated by the state, the development schemes and migrant Ittu. Accordingly, farming as risk management and livelihood strategy and the consequent land enclosure and fragmentation have eroded the traditional customary rules of land holding and led to dissecting of the best communal pasture lands which has seriously affected livestock performance. In the following sections, I explain the new forms of accessing resources among the Karrayu community, while dealing with both direct and indirect sources of vulnerability.

6.4 BRINGING NEW PRINCIPLES IN: FENCING COMMUNAL PASTURE

In the past, land in the Karrayu area was governed by customary institutions. However, with the destabilization of customary resource management methods discussed in Chapter Five, the role of these traditional institutions in the Karrayu pastoral community has declined. For instance, an increased influx of non-local farmers from other areas, particularly the Ittu groups from West Harareghe, has brought about new worldviews and resulted in the expansion of small-scale cultivation in Karrayu territory through private enclosure. As a result, the handling of the natural resource base, mainly land, has also undergone changes. In other words, with the increased social sources of vulnerability, Karrayu pastoralists have started to put in place mechanisms for handling the scarce resources, such as private enclosures, by using their 'pastoral agency'. This means, the well-off Karrayu herders with large numbers of livestock did not sit back and watch while the communal pasture lands were taken over for farming purposes. Rather, as individual agents they introduced the idea of enclosing the land for grazing purposes, either individually or in a group. Locally, land that is enclosed for private farming purposes is called *qonna*, while land that is enclosed for grazing purposes is known as *kalloo*. In other words, the engagement of the pastoralists in counterbalancing the enclosure of land for private farming has led to the introduction of new practices among the Karrayu that restrict open access to communal pasture. These were introduced among the Karrayu after the shrinkage of their resource base, and represent an attempt to actively create mechanisms for dealing with new problems. *Kalloo* is aimed at resolving the problem of shortage of pasture in the seasons of *birraa* (autumn) and *bonna* (winter). Restrictions were placed on a portion of grazing land in their permanent villages by agreements made by the pastoralists themselves through the village leader. This decision was announced to the *Woreda* administration so that it would not be violated by anyone. The portion of grazing land on which restrictions are placed remains fallow until it becomes accessible to all members of the village in the dry season by another decision of the village leaders. But nowadays the severe shortage of land caused by the migration of the Ittu pastoralists from Western Hararghe to the Fentalle area, and the use of most of the wet season grazing zone (*onna ganna*) for settlement purposes and opportunistic farming have resulted in a failure to find land that could be preserved for the dry season. During my field research, I observed attempts to enclose and reserve land for communal grazing, but still the cattle go hungry, even in the summer. So in some years, the pastoralists give up their decision by making another agreement, and what is left to be grazed for the dry season is used to satisfy immediate pasture needs in the rainy season. This makes the enclosure mechanism for using scarce resources difficult to operate smoothly. Here one may argue that even if the intent of introducing an enclosed pasture area is good, it has failed to fully meet its purpose. The increased competition for private farming land, together with the lack of adequate rainfall, even in the rainy season, have forced the local Karrayu pastoralists to cancel their agreements to maintain the enclosed pasture ahead of the dry season. This affirms the political ecology approach that acknowledges the role of the bio-physical environment in influencing human behaviour. According to this approach, the bio-physical environment also plays a role as agent by impacting human wills and behaviours (Latour, 2004). Scholars contend that the political ecology approach needs to locate social interactions within the broader framework of ecological processes and should not be trapped in the futile debate of nature/culture duality (Gezon, 2006; Latour, 2004; Zimmerer and Bassett, 2003). In their emphasis on the decisive and influential role of the bio-physical environment in influencing human-environment relations, Zimmerer and Bassett further argue that "the environment is not simply a stage or arena in which struggles over resource access and control take place" (2003: 3).

6.5 EMERGENCE OF NEW ARRANGEMENTS

The ability of actors to access and secure scarce resources affects not only their ability to increase returns but also to deal with social and climate-related stresses. In line with this argument, the Karrayu pastoralists have started to organize essential resources in order to better handle social and climate-related sources of vulnerability. Accordingly, with the loss of spaces of adaptation that constrained the traditional mechanisms of vulnerability reduction, contemporary means of handling vulnerability such as farming are being practised by the locals.

Land transfer through sharecropping is on the increase among the Karrayu community due to the arrival of the Ittu farmers from Hararghe and with the introduction of irrigation water in the villages. Before the introduction of the irrigation scheme, it was common in some villages where farming was widely practised. To explain such behaviour, the new institutional economists argue that strategic actors, acting both collectively and in their own individual self-interest, negotiate institutional structures and rules over time in order to arrive at increasingly efficient outcomes. In this sense institutions are the "by-product of (a) strategic conflict(s)" that, intentionally or not, coordinate(s) interests among disparate actors, arriving over time at equilibrium, or in other words "commonly agreed upon norms" (Ensminger and Rutten, 1990).

These are the ways through which the pastoralists attempt to utilize the limited spaces of adaptation that require different mechanisms than in the past. For instance, in the Karrayu pastoral community of Upper Awash Valley, *ye-ekul* (half share of the produce) is becoming part of the contemporary practice of handling vulnerability. Nevertheless, the big problem at the current time among the Karrayu households at the study sites such as Giddara is to find partners for sharecropping arrangements, since most of the community members want to rent out either part or all of their holdings. Fantale Jilo, who has no oxen to cultivate the land himself, expressed his situation thus:

> ...a few years back, people from town were looking for land and it was good for me. I also had an opportunity of renting my plot, which helped me to share better harvests. Now, the land does not yield much like before, and we had crop failures during successive years. As a result, half a hectare of my land is given up as fallow because I could not find someone who is interested in it.

However, the sharecropping arrangements are not always smooth for two main reasons. The first is the problem of crop failure, which according to the Karrayu households who practise farming is caused by excessive soil salinity and lack of farm inputs. The second is the gradual erosion of assets, particularly through the sale of oxen in order to meet subsistence needs. Farmers dare to hire land only if they possess sufficient draught power.

I observed another type of land transaction being practised by farmers around Galcha and Giddarra villages. Under this arrangement a farmer who is sharecropping land passes it onto a third party without the consent of the owner. One of my informants, who asked me not to reveal his name in connection with this matter because of the legality problem and the risk of losing trust and value in the community, stated:

> I sharecropped one hectare of land on the '*ye-ekul*' (equal) basis from a female-headed household in our community a few years ago, when I was able to use all necessary inputs on the farm. In the meantime, because of some constraints I encountered, I had to look to a close friend who could make up my shortfalls. This was how I managed the last two harvests. We provided half of the harvest to the landholder, and then shared the remaining produce equally between ourselves. I did not inform the individual from whom I initially got the land about the reality of the situation because of fear that she will snatch it back from me.

The commercial farming operations along the Awash River in the 1950s and the 1960s adversely affected local pastoral strategies. Pastoralists lost their entitlement to the ever expanding farms. As has been shown in Chapter Three in particular, these social problems aggravated the vulnerability of the locals to drought. In response to these multiple sources of vulnerability, the Karrayu pastoral groups have started to use their agency and take up farming as a strategy in order to deal with these vulnerabilities. Initially they pursued farming as a strategy by locating their farms adjacent to the sugar cane and fruit plantations where water is used for irrigation. They have used peaceful and less peaceful mechanisms, ranging from simply benefiting from the extra water that they get from the plantation, to damaging the infrastructure overnight and diverting the water to their own fields. Thus, they use their power to manipulate the imposed structures that render them vulnerable. This is in line with what Scott labelled 'weapons of the weak' (1985). Recently the involvement of the Karrayu pastoralists in farming activities has increased with the arrival of the new irrigation scheme by the Oromiya regional state[10].

However, the taking up of farming as an active response to the changing socio-cultural situation is not uniform among all Karrayu pastoralists. It is also necessary to ask who is actively participating in the farming enterprise in the pastoral villages as there are varied reactions to the new irrigation scheme. Not all pastoralists have reacted positively to the intervention, depending on the number of livestock they possess and the meaning they attach to the resources on the land. In the words of one of my informants in the pastoral village of Illala, who owns more than thirty camels:

> I am praying to Waqqaa so that he will break the 'legs' of the tractors. They are tilling the land on which the acacia trees used to grow. They are cutting out the trees on which our camels browse. They brought the water to our villages and now they want us to put all our livestock aside and start farming. This is impossible! [Field note in Illala Village, 2010]

For those pastoralists who own significant numbers of livestock and who wish to follow the traditional risk management and livelihood practices, the increased push by officials[11] towards farming through distribution of land in the pastoral villages is a further annihilation of their pastoral spaces of adaptation.

---

10 This scheme held up by the government as a flagship of good development. The scheme has led to the resettlement of 4,500 Karrayu pastoralist households out of an anticipated 22,000.
11 Not all government officials give full support to the conversion of the rangeland. Informal communications with few officials revealed that in view of the nature of the soil and the skills of the Karrayu, the scheme should have been based on a 'livestock major crop minor' principle. In the current conditions, they argue that politics have been prioritized over sustainability of the project.

Many of my pastoral informants appreciated the abundance of water in their villages, which was made possible by the arrival of the Fentalle irrigation scheme, as it helps them to experiment with new mechanisms for handling vulnerability. However, despite these new mechanisms for handling social as well as climate-related sources of vulnerability, the associated clearing of the remaining bushes and trees from the range land is seen by the rich pastoralists as counterproductive.

6.5.1 Benefit-sharing arrangements

One of the new institutional arrangements for accessing resources that I observed in my study areas is an arrangement in which profits are shared equally between someone who rents out land for mutual purposes and an individual willing to commit himself to a sharecropping agreement. The Karrayu with a parcel of land rents it to someone with money and oxen. The person who leases the land prepares the plot for farming by putting up resources in addition to the oxen and labour. In many cases, the person who rents the land prefers to produce cash crops such as onion, watermelon and tomatoes that can easily be marketed. Both parties share the benefits after all the costs incurred have been deducted as expenses. However, the Karrayu pastoralists do not like such arrangements, as the second party may deliberately exaggerate their investment costs. In the process of interaction with other farmers, the Karrayu have developed a profit-sharing arrangement in which they hire a farmer who works on their land, rather than renting out the land completely. This arrangement saves the Karrayu from the risk of not getting the proper profit from their land, as in the kind of arrangement where they rent out the land to someone who invests on the land. My informant in Giddarra village, Jillo Hawas, described the arrangement for me as follows:

> There are people who are interested in farming on my land because I have farm land which is near to the irrigation canal and not far from the tarmac road. So, there is no need for me to be in a hurry and rent out my land to the clever urban dwellers. They always exaggerate their costs and expenses for fertilizer and many other things. I have learned from that and I am doing it the other way around. I came to know an Ittu farmer who knows what to grow when and is good at identifying the weeds that affect the crops. So, I provided him with what he needed and he works on the farm. Then, after the harvest is sold, I take out all my expenses and then we share equally what is left as profit. There is no chance to deceive because we know each other very well and farming is not only this year. I also need this very good farmer next year. We don't want to lose trust. (Interview with Jillo Hawas, March 2010)

However, one of the challenges that the Karrayu land owners mention is the problem of initial capital to cover all the costs of farming so as to get a better result. Sometimes they may not be able to satisfy the demands that come from the knowledgeable farmers who ask them to provide them with specific inputs at a specific time. In the absence of sufficient capital they jeopardize the outcome of the harvest.

6.5.2 Resource-sharing arrangements

This is also a common kind of arrangement in the well-watered areas of the Karrayu villages. In such an arrangement, Karrayu pastoralists who have been allocated a plot of land to work on, but who do not have the sufficient start-up capital, borrow an amount of cash from a money lender. In their agreement, land is conditionally transferred from its holder to the money lender. In many circumstances, the money lender is an outsider who is involved in different forms of sharecropping. This agreement is arranged in such a way that if the borrower cannot pay the money back within the agreed period of time, the lender will take over the land owned by the Karrayu borrower until he pays the cash back. The lender in this case uses the land for various purposes according to his wishes. In some cases, poor Karrayu households enter into such contracts with their fellow rich Karrayu pastoralists. An interview with a Karrayu farmer in Illala village shows us this case clearly:

> I: How do you get the capital to buy all the inputs for your farm?
> Hawas Xinno: We borrow money for a certain time and we pay it back.
> I: Is it so easy to get the amount of money you need to operate your farm?
> Hawas Xinno: No, it is not. But what would you do if you were us? It is all about trust. Because if we cannot pay it back within the agreed period, they take the land from us! Then, we may be obliged to sell some of the limited livestock that we have to pay back our debt.
> I: So, what do you think is the solution?
> Nura Xinno: Nowadays the rich pastoralists who have many camels like to do business ('*daldala*') with us. For them it is very easy to sell one camel for 5,000 Birr and to lend us the money. You know they are also relatives. When it comes to them we know that we will not lose our land. They also give us a very flexible period and when their cattle come to the villages, they can graze them on the straw of the maize field. In that way we support each other.

In some cases, particularly in those villages that are mainly inhabited by migrant Ittu farmers, it is not easy for a Karrayu pastoralist to claim access to land and practise the different arrangements mentioned above. In these villages, cultivatable land is mainly under the ownership of the migrants who enclose a plot of land for irrigated cultivation. In the villages of Gara Dima and Giddarra I observed situations where the Ittu own and cultivate land, ranging individually from one to two hectares, on which they produce vegetables and fruits such as onions, tomatoes, watermelons, guava and also *khat* for the market. During my field research I observed many Ittu farmers entering into various forms of arrangement with the Karrayu. The demographic pressure caused by the arrival of the Ittu in the area during the 1974/75 drought has resulted in the loss of land that the Karrayu formerly used. This has a critical implication for the Karrayu, as they do not have the necessary capital to practise farming. They own only labour and capital in the form of oxen and lack land. However, despite loss of their land to the migrant Ittu farmers, the Karrayu have harmonious relations with the Ittu. This is manifested in the different symbiotic arrangements entered into by the two groups in order to pursue their livelihoods. Jarra Waday, an informant from

Giddara village, explains the different arrangements that he has with his fellow Karrayu in the following manner:

> We are practising various forms of mutual support. We need each other very much. We are intermarried. When it comes to resources, for instance, I have a Karrayu friend from whom I hire an ox to cultivate my land and produce maize. You know he has many oxen. It is not easy these days to buy an ox. You have to save for a while. Then, by the end of the harvest I give him around five to six quintals of maize in return. In some cases, these oxen are not well trained. They are scared of people and sometimes run when they hear the sound of a car from a distance. So, if my Karrayu friends want me to train their oxen for ploughing, I do that for them but free of rent during the training time which may last at least one year. The same is true for us. You know, the Karrayu herders know the best grazing sites for the cattle and I give them my milk cows to take care of them and in return I pay them with a certain amount of maize. This is how we support each other. [Field note, Giddarra Village, April 2010]

In addition to these inter-household labour- and resource-sharing arrangements, the rich Karrayu pastoralists who have to support polygamous families also benefit from the various livelihood activities for they can minimize the problem of manpower needs to engage in activities such as cultivation besides taking care of their livestock.

Table 8: The different practices and mechanisms for securing resources

Pastoralist wealth category	Engagement in cultivation	Engagement in pastoral activities[12]	Means of securing resources
Rich	Very high	Very high	Family labour, hired labour, sharecropping, renting land, cattle entrustment
Moderately rich	High	High	Family labour, renting land
Medium	High	High	Hired labour, family labour, sharecropping, renting out land
Poor	Minimal	None	Renting out land

Source: Compiled from focus group discussions

The wealthy Karrayu pastoralists try to divide their families between cultivation and herding livestock though they oppose the conversion of the communal land to farming land. According to my research informants, they do this by selecting one wife from an Ittu group who is experienced in farming, while the other wife is a Karrayu who has a very good knowledge of livestock handling. Households which are thus differentially positioned with regard to managing livelihood risks and

---

12 The scales were identified during focus group discussions, ranging from high involvement in cultivation or pastoral way of life to non-involvement.

dealing with drought hazards try to create patron-client relationships in order to secure access to resources. Furthermore, those rich Karrayu households that engage in cultivation in addition to pastoralism cope with the problem of labour shortage by employing mechanisms like cooperative labour. They try to shorten the time that they have to invest in farming tasks, like weeding, clearing or ploughing, by organizing a collective work party and providing food and *khat* for those who lend a hand. Poorer Karrayu households find it difficult to make such arrangements, and they are obliged to carry out the farming activities by themselves as they cannot afford to hire labour. In addition the shortage of agricultural inputs such as oxen makes it difficult to effectively practise farming as a livelihood strategy which would buffer them against the risk of food insecurity.

Photo 7: Land preparation by Ittu farmers

Source: Author and Oromia water works design office

However, one should not always see the poor Karrayu pastoralists as powerless victims who lack the essential resources to practice cultivation as a livelihood strategy. Rather, they use various channels to access these resources by going beyond their spatial confinement within their locality. During my stay in the field, I observed poor Karrayu farmers working on their farms with oxen that they managed to borrow from the neighbouring Arsi Oromo highland farmers. A poor Karrayu farmer in Giddarra village explained the case as follows:

> It is difficult for me to buy farm oxen to use the new irrigation scheme. Others who manage to combine oxen, labour and land are enjoying the harvest. So, I had to contact my friend who lives in Arsi. I received two oxen from him to take care of them and feed them properly while I am using them to plough my land. After using them for six months for free I returned them back. He gave the oxen to me because the availability of fodder here is better than in his area. But, for me it means a lot. After selling part of my produce and some goats, I have managed to buy my own ox (November 2011).

These various mechanisms and strategies that are employed by the Karrayu pastoralists in order to access critical resources at times when their social and environmental set-ups are fragmented imply the practical-evaluative element of pastoral agency in transforming their surroundings despite structurally imposed constraints (Giddens 1976; Berry 1989; Emirbayer and Mische, 1998). Furthermore, actors can effectively obtain access to resources through a variety of 'strategic acts' such as exchange, gifting, and claims (Ribot and Peluso, 2003). As has been demonstrated by the way the Karrayu pastoralists resort to a variety of sharing arrangements, access to resources at times requires engagement in a range of social relations, including those of the market. In doing so, actors must be able to navigate a dense, overlapping institutional landscape, a nested hierarchy of authority and control, ranging from formal powers such as the state to informal ones like heads of households, which differentially distribute channels of access or entitlement (Berry, 1993; Leach et al., 1999; Ribot, 1998). As in the case of the rich Karrayu pastoralists, some actors are better situated than others to achieve such goals, but this does not always prevent actors from trying to manipulate powers of authority and control to their own ends. In other words, in the case of the Karrayu individuals who practise farming as an adaptation to livelihood insecurity, obtaining access to resources is encapsulated in a dialectic process of negotiation and bargaining among various forces and powers at different scales of social interaction. For the Karrayu pastoral households, the conditions on which they formerly relied to access their resources have been distorted by various factors and they have responded accordingly. This clearly shows us that at times when the conditions that ensure access to resources change and actors are faced with uncertainty, they must actively seek other solutions that will enable them to realize their goals (Leach et al., 1999).

6.6 SOME CONSTRAINTS TO CULTIVATION AS A CONTEMPORARY STRATEGY

Karrayu households who are engaged in farming practices have been facing a number of challenges to crop production, for which they have developed their own coping mechanisms and strategies. Some solutions are connected with so-called 'modern' practices. As can be seen from the table below, farmers are aware of most of the problems related to their farm plots and have their own short-erm and long-term land management techniques to minimize the impacts on crop production. Households apply various forms of indigenous knowledge to minimize the risk of crop failure and drastic declines in yields.

Soil fertility improvement techniques: The number of years under continuous cultivation and crop yield decline are good indicators and are mostly used by the farmers to judge the soil fertility levels of farm plots. They attempt to maintain and restore soil fertility mainly through traditional mechanisms, and recently a few households have been trying to adopt modern methods. The main traditional methods include crop-rotation, application of manure, intercropping, and fallowing, which were reported to be the major response strategies employed by the households practising farming as a livelihood activity. The possibility of resting land from cultivation has become very limited because of land scarcity. Thus, fallowing is almost non-existent at the study villages nowadays. The elderly inhabitants at the study sites were able to recall the situation in the past when they used to give up land after two consecutive harvests, a practice which was rather similar to shifting cultivation. The use of inorganic fertilizers to replace soil fertility loss was practised by very few farmers at the study sites. Those Karrayu farmers who reported using fertilizer informed me that they have to rely on the knowledge of the neighbouring Tulema Oromo[13] farmers who have experience in using fertilizers for their fields.

Table 9: Farming households' responses to constraints on crop production

Constraints on Crop Production	Households' Responses
Soil erosion	Planting trees with annual crops
Poor soil fertility	Animal manure, crop-rotation and fallowing
Drought (water scarcity)	Irrigation, change in cropping pattern
Pests and insects	Using insecticides and pesticides

Source: Compiled from interviews with households at the study sites

The different households in Fentalle *Woreda* have extensive experience of catastrophic droughts and how they affect the environment, as well as human beings and their activities. The farmers have learned how to cope by using short-term and long-term survival strategies. With regard to crop production, there has been a clear shift from emphasis on subsistence crops to emphasis on cash crops, as well as multi-cropping, which can be considered as a response mechanism to

---

13 The Tulema Oromo are groups who reside in the neighbouring Arsi zone of Oromiya Regional state and who are mainly engaged in farming as a livelihood strategy.

drought. Above all, utilizing irrigation for crop production, individually or cooperatively, is the best coping mechanism to overcome the risk of crop failure caused by water scarcity. The big problem facing the farmers at Giddarra and Illala villages in particular, as well as the rest of the villages in the district, and one that remains as the main challenge to the Development Agents working in the area, is a lack of awareness regarding the excessive accumulation of salts in the soil. The invasion of a notorious weed that is locally called *shiferaw* (*Parthenium hysterophorus*) is also spoiling the farming lands, a problem for which both the locals and the agricultural officers do not have a solution thus far. Crop yields have drastically declined over time, even with sufficient water through irrigation. In the words of an informant in Giddarra village:

> I planted maize two years back using irrigated water and got a good harvest. Last year I planted the same and it was a total loss. As you see water is not a problem. But it is the land that says 'I don't know you' as if it is not your land and will not give you a positive harvest. [Jarra Waday, April 2010]

The problem of access to markets: Farming as a livelihood strategy among the Karrayu households is becoming increasingly important as an adaptive response for those who are engaged in it. The significance of farming as a source of cash is also high as it helps the households in covering some of their expenses, such as school fees for their children and clothing. However, this activity as a livelihood strategy is not without risks. One important factor that hampers this activity is the problem of marketing. Due to their lack of experience and orientation, almost all Karrayu pastoralists produce the same cash crops at the same time using irrigation. This has created an excess supply in the market, as I was able to observe during my stay in the field in 2011. Nuri Qumbi of Illala village explained for me what he experienced:

> The development agents did not advise us what to produce when. They just distributed the land to us and told us to farm. As we used to observe our friends selling a kilo of onion for 6 Birr in other villages, and as we thought that we could get cash immediately, we planted onion, everyone planted onion. Very few planted maize. The harvest was incredible. But, the merchants came to our village and told us that onion is now all over the country and they wanted to take our onions for 0.50 cent per kilo. Imagine how much labour, money, and time I spent on it! I would have said no to their offer. But, you cannot put your onions in a granary [Field note, November, 2011]

6.7 NON-PASTORAL AND NON-AGRICULTURAL ACTIVITIES

In their study, Barrett, Reardon and Webb (2001) make a distinction between three forms of non-agricultural activities: local wage labour employment, local self-employment, and activities undertaken following migration to other areas. The importance of these activities in the livelihoods of rural populations in the developing regions has been growing fast. The corresponding figure for Ethiopia is 36% (Reardon 1997). This increasing trend in the proportion of people making their livelihood from non-farm activities and the shift from agriculture to non-agricultural ventures is defined by Bryceson (1996) as 'de-agrarianization'. She explains it as a long process involving three important aspects. First, it entails 'economic activity reorientation' or the change in main sources of livelihood, whereby people start to generate income from non-agricultural activities. Second, it means an 'occupational adjustment', whereby a certain segment of the labour force moves out of agriculture. Third, the change in type of activities is accompanied by a change in land use patterns (1996: 99). My observations from the field show that a substantial number of Karrayu pastoralists are engaged in non-pastoral and non-agricultural activities such as wage labour and sale of charcoal to generate income. A brief account of the main non-pastoral and non-agricultural sources of livelihood is presented in the following section.

Wage employment
The establishment of commercial farms in the Karrayu area created employment opportunities. Although the irrigated farm schemes created employment for large numbers of temporary and permanent workers, it was not the displaced Karrayu that benefited from these job opportunities. Although the Metehara sugar factory has a high demand for labourers to perform tasks like cutting sugar cane, land preparation and irrigating the fields, the Karrayu pastoralists are unsuited for such work due to their pastoral background. However, in recent times Karrayu have been employed by the sugar factory in various departments (see the table below). This is part of the increased diversification strategies that the Karrayu employ. According to the Metehara sugar factory personnel department, there are more than 900 Karrayu individuals who are currently employed by the factory, both as skilled and unskilled labour.

Table 10: Number of Karrayu individuals employed in the MSF

Department/Section	Type of employment		
	Permanent	Seasonal	Contractual
Cane harvest	15	63	48
Garage & Maintenance	7	7	–
Land preparation	1	1	–
Plantation	5	628	192
Administration	1	–	–
Production	18	–	–
Total	**47**	**699**	**240**

Source: MSF human resource department, 2012

Firewood selling as a subsistence-oriented activity
Firewood selling is the main option for female-headed households, although members of male-headed households are also involved in this venture. Next to wage labour, poor pastoralists rely on this activity. What makes it different from other income-generating activities is its relative openness to those who are capable of undertaking the collection and transport of the material to market places. I was told that the collection of wood takes one day and taking it to market is done on another day. There are two difficulties met by the people attempting to make a livelihood from this activity. First, the price of firewood goes down when a large number of community members are involved in selling it. Second, the sources of wood are becoming very inaccessible due to the dwindling natural resource base in the nearby villages. This subsistence-oriented activity has long-

term implications for the environment. However, one of my informants in Illala village commented:

> When there is no food to eat in the house and the children are hungry, how can I have peace of mind? In a situation where we lack daily supplies, I do whatever is readily available. Who has the time to think about the future? [Illala village, 2010]

Photo 8: Selling of fuelwood and charcoal in Metehara town

Source: Author

The above statement shows that for poor pastoralists a livelihood problem often impacts the existing support system. Selling charcoal and firewood is the most common activity recorded among poor Karrayu pastoralists. However, these ways of generating income should not be seen as chosen or preferred, but instead as "last resort" options adopted by people who are poor and desperate for any income at all.

6.8 SUMMARY

This chapter has explored the contemporary risk management and livelihood practices pursued by the Karrayu pastoralists in Upper Awash Valley using their evaluative-practical and projective agency. These mechanisms for re-organizing resources are responses to the socio-cultural sources of vulnerability rather than reactions to climatic-related stresses alone. It has been discussed in this chapter that among the major socio-cultural sources of vulnerability it is population pressure as a result of the Ittu migrations that has contributed most to the emergence of new forms of accessing scarce resources by the Karrayu pastoralists in Fentalle *Woreda*. However, the migrations of the Ittu farmers have also brought the transfer of skills and knowledge into the Karrayu community, thereby helping the pastoralists to develop new worldviews with respect to handling vulnerability. Individuals, households and communities are active agents that make choices based on opportunities and anticipated outcomes; however, these are framed within a power structure of social and political relation (Eriksen & Lind, 2009). One should not assume that pastoralists are passive agents in the face of the changing contexts discussed in Chapter Four. Pastoral communities and households have taken up agro-pastoralism as a rational livelihood diversification and risk management strategy.

In response to these socio-environmental changes, the Karrayu actively experiment with new ways of managing their scarce resources, thereby exercising their agency to deal with the structural constraints they have been confronted with. In the process, the traditional mechanisms of handling vulnerability have given way to new mechanisms of vulnerability reduction. The decline of livestock-based vulnerability reduction tools also means less cohesion of these tools in the social fabric of the pastoral community, rendering the locals more vulnerable to climate stresses. For instance, as has been shown in this chapter, the process of fencing off communal grazing land for private farming purposes has implications for the re-organization and differentiation of Karrayu pastoralists, as it means different things for the different pastoral groups depending on their wealth status. These different worldviews also have implications for the climate-related vulnerability of the Karrayu pastoralists. For instance, contemporary practices such as contractual arrangements in respect of labour and other critical resources such as land and livestock, have important implications for the meaning that Karrayu households attach to resources and the ways in which certain people have become positioned within the broader cultural fabric. In this chapter I have tried to show the connection between pastoral agency through new mechanisms of dealing with vulnerability and the socio-cultural and climatic sources of vulnerability. Rather than the loudly drummed 'climate change impact leading to adaptation' premise, our observations among the Karrayu pastoral groups in a dry-land context reveal 'social vulnerability leading to changes in adaptation strategies'. In the following chapter, I reflect on the dominant climate change adaptation discourses in the light of the local level processes of vulnerability and adaptation, so as to show the relavance of locating adaptation to climate-related stresses within specific socio-political contexts.

7. CONTEXTUALIZED ADAPTATION: HEGEMONIC PERSPECTIVES AND LOCAL RESPONSES

"Climate change is both a narrative and material phenomenon. In so being, understanding climate change requires broad conceptualizations that incorporate multiple voices and recognize the agency of vulnerable populations."

Farbotko and Lazrus

7.1 INTRODUCTION

As set out at the beginning of this book, the hegemonic perspective on adaptation to climate change is not critical enough in locating climate-related risks in their respective broader Socio-political contexts. For instance, the Intergovernmental Panel on Climate Change (IPCC) defines adaptation to climate change as "the process of adjustment to actual or expected climate and its effects [and] to moderate or avoid harm or exploit beneficial opportunities" (IPCC, 2014: 5). This narrow definition of adaptation focuses on tackling climate change impacts in separation from broader contextual stressors – social, political, economic and cultural (Adger, *etal.*, 2009). This type of adaptation requires changes based on climate predictions.

In line with an adaptation perspective that emphasizes understanding the social and political contexts, the three main empirical chapters, on the historical context of livelihood insecurity, pastoralists' livestock-based risk management and livelihood practices, and the move towards agro-pastoralism, show how climate stresses are situated in broader social fabrics that influence each other in shaping local level responses. Within the broader framework of political ecology with differentially located actors, I have also shown how pastoralists' adaptation is mediated by power differentials, by the actors' knowledge and their ability to strategically respond to both constraints and opportunities emanating from the changing environment. This chapter addresses the fourth specific research question: 'How does research on context-specific risk management and livelihood practices enable us to better reframe the mainstream perspective on adaptation?'

7.2 THE POLITICS OF ADAPTATION: TOP-DOWN APPROACHES TO DEVELOPMENT

Unlike the dominant understanding of adaptation that focuses on climate stimuli, the discussion of Karrayu pastoralists' livelihood insecurity in Chapter Four shows that adaptation processes are highly political and need to be seen in their historical context. Climate-related risks affect the livelihood of Karrayu pastoralists. However, other socio-political stressors, such as policy implementations that hinder existing adaptive capacity, will increase vulnerability to climate stress. These "other" stressors have their roots in the historically unbalanced pastoralists-state interactions. As argued in Chapter Four, the loss of key grazing sites has increased pastoral vulnerability to climate change. This has jeopardized the pastoralists' risk management strategies for dealing with drought when it occurs. Though the current Ethiopian government has initiated the Climate Resilient Green Economy (CRGE) strategy, its overarching adaptation strategies and the implementation of subsequent adaptation projects in the pastoral areas of the country have undermined the pastoral way of life. Many of the projects that are meant to improve the adaptation capacity of the pastoralists promote sedentarization and crop-cultivation. These interventions are highly centralized and top-down, and slowly erode existing local adaptation strategies. One of the objectives behind the promotion of sedentarization in pastoral areas is to improve rural livelihoods and conditions through modernization and industrialization measures that favour agricultural-led development (Scott, 1998). There is an idea that well-structured 'modern' communities are more productive than unstructured scattered groups. Additionally, the interest of the state is to gain bureaucratic control over dispersed and 'disorderly' pastoral populations, land and resources (Scott, 1998), while increased government control is seen by pastoralists as a threat to their risk management and livelihood practices.

7.2.1 Consideration of governance structures in adaptation

Unequal allocation of resources, adverse development policies and other national economic and social structures can influence vulnerability and hence the risk management and livelihood security of the local community (Eriksen *et al.*, 2011; Eriksen and Lind, 2009). In this way, the political economy plays a central role in shaping adaptation avenues. As shown in Chapter Four, the historical trajectories of insecurity are the result of interactions between the political economy of the various Ethiopian regimes and the pastoralists and their environment. The political economy of a country determines the processes of political decision-making (Nelson and Finan, 2009). Some forms of governance inhibit community participation and adaptive capacity at local levels (Nelson and Finan, 2009). For instance, the way the Ethiopian government is trying to transform and modernize the Karrayu pastoralists' livelihood is seriously hampering their livestock-based risk management and livelihood practices. As it is discussed in Chapter four, in the current government's policy emphasis is given to crop-based farming system that does not properly goes with the pastoral ecosystem and giving marginal attention to the livestock sector. Although there appears to be development policy and strategy to address and support pastoral production system, observations in the field show that training and extension services target crop-based livelihood through establishment of Farmers' training centres. Infrastructural support for livestock production also ranges from minimal to non-existent.

In order to counterbalance the dominance of the government and ensure proper adaptation at a local level, changes need to be initiated in the power structures that will put the vulnerable pastoralists at the centre. As shown in the previous chapters, it is the historical exploitative relation between pastoralists and the Ethiopian state that has put the Karrayu pastoralists' relations with their environment out of balance. Thus, the hegemonic adaptation discourse that focuses solely on climatic sources of vulnerability also needs to consider the broader governance contexts in which the locals develop their adaptation strategies. In order for vulnerable communities to challenge dominant power structures, their active participation in decision-making processes that affect their own livelihood is essential. Empowering communities in adaptation processes thus means a shift of power away from policy makers to local populations, which enables people to choose adaptation practices themselves instead of following policy makers' recommendations (Eriksen and Lind, 2009). Participation means a shift of power from the powerful to the less powerful, and participation can become a power struggle where participation eventually ends up as a policy narrative that does not correspond to practice at local levels. Although their intent- ions may be good, as observed in the case of the Fentalle integrated irrigation project, government officials use participation as a buzzword to legitimate their actions rather than to actually carry it out in practice (Cornwall and Brock, 2006). In these instances, regional and local government authorities control and direct participation to serve their own interests and in practice the pastoralists are not involved in the decision-making process. Participation is thus not only about physically involving individuals in the development process, but also about allowing them to define and characterize their own problems and solutions and giving them power in planning processes (Nelson and Finan, 2009). Community-based adaptation (CBA) can be one way to successfully integrate participation into adaptation approaches (Reid et al., 2009). Community-based adaptation, like any other project seeks to prioritize community needs, interests, knowledge and capacity in order to help people better cope with the impacts of climate change. Community-based adaptation addresses the potential impacts of climate change and builds resilience by integrating risk awareness from both local knowledge and scientific knowledge into its activities (Ensor and Berger, 2009). Through genuine participation, individuals and communities can be inspired to be self-reflexive on their own behaviour. Furthermore, participation also recognizes the strength of local informal institutions that could be the building blocks of adaptation.

In the process of implementing projects that are meant to improve pastoralists' adaptation to climate-related risks, it is important to recognize the structures in power relations that are being used as a medium to implement the projects. The Ethiopian government, by simply buying into the hegemonic discourse of adaptation, has implemented large-scale sedentarization projects in a top-down manner, without consulting the locals. This highly modernist top-down approach to implementing sedentarization projects is based on a one-size-fits-all idea of promoting development and improving productivity. The major reason behind the failure of sedentarization programmes in the 1970s was the lack of acknowledgement of place-based locally generated knowledge (Scott, 1998). Although some pastoralists had knowledge of and familiarity with opportunistic small-scale maize cultivation prior to settlement, this does not necessarily translate into effective sedentary irrigated agricultural production. Similarly, the knowledge produced through many years of experience on how to manage resources through livestock-based mobile risk management and livelihood practices may no longer be fitting or adequate in sedentary communities. Differentials in power relations dictate which objectives are prioritized and promoted, and create winners and losers in the process of adaptation (Adger *et al.*, 2003). Adaptation projects that propagate the ideas and interests of powerful actors will not support sustainable adaptation, but will continue to marginalize already vulnerable groups.

7.3 SITUATED AND LOCAL-LEVEL ADAPTATION PRACTICES

As has been shown in the previous chapters, the way the Karrayu pastoralists in the Metehara plain comprehend and deal with climate-related stresses is part and parcel of socio-political processes of change. The nature of vulnerability and resilience to climate stress among the Karrayu community of Upper Awash Valley is grounded firmly in factors and processes indirectly related to climate stress. My informants primarily framed their vulnerability as an outcome of various social, cultural, economic and political factors and processes, rather than as an outcome of merely climatic or environmental processes. In the perspective of the locals therefore, vulnerability to climate stress is primarily a product of 'non-climate' factors and processes. In other words, climate-related stresses often reveal deep-seated problems that have already been created by ill-planned development interventions which, for example, restrict the locals' access to key resources, as has been shown in Chapter Four. Among the Karrayu pastoral groups and others who dwell in the Awash valley, extreme climatic events in the form of drought have revealed the magnitude of other stresses, such as dwindling local knowledge and traditions, loss of customary leadership, unsustainable resource utilization, and population growth, among many others. With regard to their experience of climate stress in the form of drought, almost all the informants swiftly mentioned concerns related to socio-cultural change and the implications of non-climatic forces in giving shape to the vulnerability that arises out of the climate stress. Local cultural frames in the Karrayu community do not regard climate stresses and uncertainties as a 'force of nature' that departs from 'normality'. As discussed in Chapter Five, dealing with environmental uncertainty is – and has been over generations – an ingrained part of social, cultural and livelihood systems in Fentalle *Woreda* of Upper Awash Valley. In the words of my key informant, Abba Gada Kawoo,

> ...for us drought was not a problem...it comes in the dry season and we go to the wet grazing pasture. We also keep more animals and if it takes some of them then we continue with the rest. We knew what to do during the dry and wet seasons. For us drought was like a signal that tells us that now we should move to the riverside under the shade of the big trees...that tells us that the grasses along the river are ready to be used...so drought was not a problem at all.

In line with the above statement and as has been shown in the empirical chapters, in the understanding of the local groups in Upper Awash Valley, the natural environment is ontologically combined with broader cultural, social, political and economic features. This way of thinking differs from the dichotomies inherent in Western thought. In the Karrayu pastoral community, 'the environment' includes culture, identity, economy and politics. Accordingly, my informants understand the influence of extreme climatic events such as drought in tandem with the influence of socio-cultural factors, and see a close connection between climate-related stresses and the concerns and opportunities of everyday life. In the following table, I summarize the local and the government constructions of the climate stress problem and the solutions forwarded. It clearly reveals that the dominant 'impact reduction' orientation is based on standardized scientific solutions and undermines the social dimension of vulnerability that aggravates the problem of drought.

The Karrayu pastoral groups in the Upper Awash Valley of Ethiopia understand disasters *(saxxilum)* as "extensions of the problems faced in 'normal' or 'daily' life". According to the perspective of my informants, there is little separation between the risks, stresses and problems arising in the normal course of everyday life, and vulnerability to specific biophysical climate stresses. The social structural forces behind vulnerability to climate events, such as loss of grazing land to commercial farms and conservation areas, socio-cultural change and population growth, were mentioned by the Karrayu pastoralists. These social sources of vulnerability are of priority concern to the community, irrespective of climate stress. The contexts of livelihood insecurity are discussed in Chapter Four as they have more far-reaching implications for the pastoral way of life in the semi-arid areas of upper Awash Valley than climate-related problems. 'Non-climate' stresses are an everyday concern because they limit the ability of the pastoralists to exercise their rights and maintain their values. The social sources of vulnerability that are discussed in Chapter Four have the effect of narrowing down the spaces of adaptation that the Karrayu pastoralists used to rely on in the past to maintain a socially acceptable way of life based on livestock keeping.

7.3.1 Recognition of local agency

Despite the prevailing apocalyptic discourse on pastoralists in relation to climate change, the local community use their agency and deal with vulnerability arising from both social and climate-related sources. In the process of their interaction with government projects to promote sedentarization, the pastoralists are not passive. Communities and households among the Karrayu have taken up agro-pastoralism as a rational livelihood diversification and risk management strategy. Individuals, households and communities are active agents that make choices based on opportunities and anticipated outcomes; however, these are framed within a power structure of social and political relation (Eriksen and Lind, 2009). As has been discussed in both the theoretical and the empirical chapters of this dissertation, human agency is the capacity of human beings to make and exercise choices within their own cultural milieu. The Karrayu pastoralists have been exercising their nomadic way of life within the limits of the resources in a dryland environment for centuries. In other words, drought has been part and parcel of the traditional social fabric to which they have developed response strategies over the years. Hence, ways of dealing with climate-related stresses are not seen separately from other everyday social practices.

By contrast, the 'naturalization' of climate change impact in the dryland areas of Ethiopia has led to interventions that undermine this pastoral agency and resilience. In the eyes of the Karrayu pastoralists, it is the loss of key watering points and grazing land for their animals that has made them vulnerable, far more than the fluctuating rainfall which is a familiar feature of this arid environment. They see climate-related extreme events, such as drought, in relation to other social factors that influence their ability to access water and pasture.

Even though the traditional pastoral responses are undergoing changes as a result of various sources of livelihood insecurity, it has been shown in Chapters Four and Five that the Karrayu pastoralists are active agents in addressing their own identified climate-related problems. Many of the new mechanisms for handling vulnerability explained in Chapter Six show that the Karrayu pastoralists are by no means passive victims of climate-related stresses. However, many of the development interventions driven by 'political' agendas in the Karrayu pastoral community hamper this agency by prioritizing farming over pastoral activity. These interventions ignore the importance of the skills and know-how of the pastoralists themselves, and associating pastoral adaptation with a shift to farming will affect the resilience of the locals in the face of climate stress. In the words of one pastoralist from Illala village:

> I don't know what they want us to do. They bring the irrigation and they told us to farm. Why should we farm? It's better for us to keep our animals. When we refuse they lock us in this big warehouse for one or two days. They also tell the elders that they should 'correct' us. They think that we are lazy because we refuse to farm. They think that we lose more livestock simply because we are pastoralists [Roba Bultom, Illala, 2010].

The above statement shows that local pastoral agency has been excluded from development attempts that are meant to increase the response capacity of the locals by providing access to critical resources such as water. However, access to irrigated water does not necessarily guarantee the exercise of power by the pastoralists over these resources, or enable them to use their agency towards attaining their own objectives, which are centred on raising livestock. For instance, it was commonly said by the pastoralists who participated in this research project that they do not have any grudges against the irrigation scheme being implemented by the regional government, for it helps them to produce fodder for their animals. This, together with their innovative way of fencing pasture land on a group basis, is an important part of their everyday efforts to deal with the climate-related stresses that affect their livestock.

Thus, contestations over the utilization of limited spaces for a particular vulnerability-handling strategy have led to the inscription of various responses into the Karrayu pastoralists' adaptation space. As has been shown in Chapters Five and Six, the Karrayu are using their iterational and practical-evaluative agency to maintain their pastoral activity, while at the same time engaging themselves in active appropriation of resources from the 'spaces of farming' through arrangements such as sharecropping. This shows that pastoral actors, in order to secure access to resources necessary to their livelihood goals and strategies, attempt to utilize existing institutions, draw on past ones, or create new ones, in order to channel social action so that outcomes can be expected with a greater degree of certainty. In this political process pastoralists mobilize their assets, both economic and social, while weighing costs and benefits of potential courses of action, in order to further their strategies, manipulate political alignments and social configurations, and accomplish their ends.

All in all, this process of mixing strategies may not necessarily mean diversification. Rather, the simultaneous and active involvement of the Karrayu pastoralists in several strategies represents what can be considered as 'hybridization'. Hybridization denotes reflexive awareness of imposed and introduced elements. In their attempt to deal with the ever expanding practice of farming while keeping their pastoral way of life, the Karrayu create mechanisms that can help them handle both social and climate-related sources of vulnerability. The creation of these hybrid arrangements reveals pastoralists' creativity in their encounter with an alien realm or discourse, through applying wide-ranging and innovative strategies.

The participants in my research project were fully aware of the multiple sources of livelihood insecurity (Smit and Wandel, 2006). They were aware of the ways in which distant, structural factors and processes influenced and shaped many local problems, both climate-related and 'non-climate' related. In terms of event-centered vulnerability, structural processes were identified by the participants as the primary and underlying causes of reduced 'choice' in responding to climate stress, in terms of both decline in traditional vulnerability-handling mechanisms and limited effective contemporary risk management and livelihood strategies.

In terms of the PAR model, the Karrayu pastoralists have a broad understanding of the 'root causes' and 'dynamic pressures' that create 'unsafe conditions' and therefore shape their vulnerability in the context of climate stress (Wisner *et al.*, 2004). A participant from Giddara village attributed many problems in the community to modernity and the rapid changes sweeping through the pastoral villages, and the particular consequences of these in the Karrayu context.

As has been shown in Chapter Four, the pastoralists' traditional mechanisms of risk management and livelihood practices have declined due to social sources of vulnerability that are outside of the locals' control. My research participants emphasized that their vulnerability arises from these structural forces within the broader social contexts that have shaped their current local context and made pastoral practices more vulnerable to climate stress. The key informant elders also provided a complex explanation for their increased vulnerability to climate stress by focusing on the social context of their everyday pastoral life. In Chapter Two, I have explained that the vulnerability arising from climate-related stresses is a politico-ecological problem. The common perspective of my research participants is that many of their traditional mechanisms for handling climate-related vulnerability have become increasingly dissociated from the practices, structures and worldviews of their everyday life. The empirical chapters have clearly shown that the perspective of the Karrayu pastoralists reflects the perspective of vulner- ability as used in the field of disaster research rather than climate change adaptation. Hence, tackling the challenges faced by them in their everyday life is tantamount to reducing their vulnerability to climate stress.

Table 11: The pastoralists-government understanding of adaptation

	Seeing like the Karrayu	**Seeing like Ethiopia/Oromiya[14]**
Construction of drought	Drought is part of everyday experience	Drought 'happens to' the Karrayu pastoralists
Sources of vulnerability	Drought is a problem due to loss grazing land and pasture	Drought is a purely meteorological problem
Handling vulnerability	Traditional mechanisms of vulnerability reduction, such as mobility; practices of resource management are important in reducing drought-related livestock loss	Sedentarization along the river bank and farming are the only way out; tackling impact of drought
Assumption behind responses	Differentiated strategies in handling vulnerability, depending on the position of individual pastoralists	Fast forward planning and projects based on the assumption that one size fits all
Meanings Behind resources	Access to irrigated water and entitlement to the Awash river is good, but 'let livestock and our skill and knowledge' be at the centre of a better life	Large-scale investment in diverting the Awash river to pastoral villages, and then the Karrayu have to settle and start farming and keep few livestock/km^2

[14] The government perspectives are compiled from official regional and federal government documents, such as Fentalle integrated irrigation project documents and CRGE.

7.4 SUMMARY

By considering the three empirical chapters together in the light of the theoretical orientation of this dissertation, this chapter has synthesized the conceptual and empirical analyses contained in this study. On the basis of insights from the Karrayu pastoral community of Upper Awash Valley, it has been shown that local level adaptation in a semi-arid environment requires the consideration of 'pastoral agency' and the inclusion of initiatives that address the development-related causes of increasing vulnerability. The types of activities that this would involve are not directly related to climate stress or climate change. In other words, local level adaptation to climate-related risks are part and parcel of the social fabric of the pastoralist community, and hence development interventions at a local level have to be tailored towards building on the foundation of the locals' resilience, rather than focusing on isolated new interventions for adaptation. One way to increase the local pastoralists' resilience is through the revival of some of their traditional tools for managing resources. There are already instances where the local people themselves are trying to revive the old mechanisms for managing the range land. Empowering such initiatives by the locals themselves and increasing the ability of pastoralists to live in a modern world is crucial in dealing with vulnerability to climate-related stresses. Through local eyes, vulnerability is clearly a politico-ecological process. Hence, the betterment of local level adaptation eventually entails transformations in power structures at the national, regional and local scales.

Adaptation interventions at the local level should consider the social conditions that perpetuate vulnerability. Among the Karrayu pastoral groups in Fentalle *Woreda*, social sources of vulnerability precede, and are more relevant to, the locals than climate-related stresses. People practise livestock mobility and keep a significant number of livestock varieties which can help them better utilize their resources in the semi-arid environment. However, vulnerability is increasing because of primarily social factors. In this respect, any intervention attempting to reduce vulnerability among the Karrayu pastoral community should not wait until climate change becomes high on the agenda of the locals. Implementing interventions meant to reduce vulnerability at a local level is a challenging task, especially when vulnerability to climate change is constructed as mainly a biophysical, environmental problem. In many cases this has led to 'pastoral areas littered with failed interventions' because they are reactive responses to specific climate impacts. Interventions that are meant to support the local pastoralists have to go beyond this reactive response and accommodate activities that are only indirectly related to climate change.

8. CONCLUSIONS

In this dissertation, I have critically explored the theoretical underpinnings of hegemonic adaptation research in relation to local level responses by locating the research problem in a specific context – among the Karrayu pastoral community in dryland areas of Upper Awash Valley. I set out to answer the question: *how have state interventions and the associated environmental transformations been experienced and acted upon by the local pastoral communities in the arid and semi-arid Metehara plains of Upper Awash Valley, Ethiopia?*

In order to answer the general question posed above, I developed four specific interrelated questions. In answering these research questions, I had four objectives. These objectives were: to explore the major social and climate-related sources of insecurity that the local Karrayu pastoralists have experienced; to bring to light the traditional livestock-based pastoral mechanisms of risk management and livelihood practices; to understand new mechanisms of risk management and livelihood practices that the locals have developed using their pastoral agency under severe structural constraints and changing environmental conditions; and to explore the relevance of context specific research on risk management and livelihood practices in order to better understand the mainstream climate change adaptation perspective, in particular its conceptual framework of vulnerability. I finish by discussing policy implications of this research and some possible directions for future research.

8.1 STARTING-POINT VULNERABILITY: THE POLITICAL ECOLOGY OF LOCAL ADAPTATION

It has been shown in the theoretical chapter of this dissertation (Chapter Two) that the widely held international assumption concerning adaptation to climate change is based on the premise that 'adaptation to climate change impacts can reduce vulnerability'. In the mainstream adaptation discourse, vulnerability is constructed as primarily a function of specific climate stimuli, their biophysical impacts, and the ability to directly respond to these. According to this perspective, climate stimuli are part and parcel of vulnerability, and minimizing the impacts of climate stimuli will minimize vulnerability to climate change. This line of reasoning puts climate change at the centre of the problem. However, I have criticized this approach to adaptation research as it focuses on actual and future climate change impacts and there is no room to accommodate other social stressors.

In contrast to the international hegemonic climate change adaptation perspective, I have based my argument on local level adaptation from the perspective that grasps vulnerability to climate change as related to broader environmental and social situations, rather than mere climate change impacts. Based on this assumption, unlike conventional research on adaptation that emphasizes climate-related hazards, in this dissertation I have shown that the Karrayu pastoralists make various flexible responses to climate-related vulnerability, using their agency and under the influence of social and demographic factors. In this regard, the research attained its objective of explicating the multiple sources of livelihood insecurity that the Karrayu pastoralists in dryland areas of Upper Awash Valley have to deal with, depending, among other factors, on geographical location, interaction with other groups, gender and generational differences. Such an understanding challenges the conventional view of seeing local adaptation as both a response to climate change and a linear process towards a particular goal. The presentations in the empirical chapters of the thesis reveal the utilization of a variety of risk management and livelihood practices, despite a seemingly homogeneous pastoral community. Although the insights and observations are specific to the study area, they nevertheless form bodies of knowledge on the political ecology of local adaptation, giving emphasis to actors, power and decision making processes.

It has been generally argued that a multiplicity of natural, social, political and demographic sources of insecurity are responsible for the way the locals interact with their natural environment and pursue their adaptation strategies. However, it is necessary to clarify which sources of insecurity are the most important, in order to explain the risk management and livelihood practices of the locals, which in turn affect the environment among the Karrayu pastoral communities at any given period. The exploration in Chapter Four showed that the Karrayu have been vulnerable to various social and climatic factors in different periods over the past six to seven decades. These sources of insecurity that have affected the entire community have been examined using participatory and qualitative approaches.

The principal sources of livelihood insecurity during the 1960s were the arrival and establishment of large-scale commercial farms and protected areas which jeopardized the local adaptation practices through land entitlement failure. In the 1970s and 1980s drought visited the valley on several occasions, seriously affecting the livelihood base of the Karrayu and rendering them more vulnerable. After the 1980s the influx of non-local migrants to the Karrayu territory and the proliferation of farming further exacerbated the problem of livestock-based risk management and livelihood practices for the Karrayu pastoralists and disrupted their social fabric and resources. All in all, it is the combination of social and climate-related sources of vulnerability that have shaped the current flexible practices of the Karrayu pastoralists. My observation negates the dominant international discourse on adaptation, which puts emphasis on climate impacts as the sole source of adaptation.

I have argued that contemporary livelihood insecurity is mainly the outcome of human failures. However, I am not arguing that the climatic source of vulnerability is not affecting the locals. Rather, the observations here are in line with the argument of political ecology that considers imbalances of power, or unequal relations between different actors, in explaining the interface between society and the environment (Bryant and Bailey, 1997). The central conclusion of this dissertation is that the livelihood insecurity of the Karrayu pastoralists and their local level risk management and livelihood practices are shaped by structural, historical processes. However, climatic hazards, specifically droughts, have very often been blamed, as though they were the sole cause of pastoral adaptation problems experienced by the Karrayu. Such an understanding is usually influenced by the 'realist perspective on hazards and risks' that views hazards not as socio-ecological phenomena but as events that happen to society. The relationship between drought and livestock rearing is quite clear. The argument is that in the past the Karrayu pastoralists implemented strategies of seasonal mobility to manage the impacts of drought on their livelihoods. However, such traditional mechanisms for handling vulnerability have been disrupted due to the denial of access to critical resources by the arrival of large-scale development interventions, and later high population in-migration that has affected the command and control of resources by the Karrayu pastoralists.

8.2 LOCATING AGENCY IN POLITICAL ECOLOGY

As has been shown throughout this dissertation, the broader social and climatic sources of vulnerability influence the knowledge, decisions, and actions of pastoralists in their attempt to access and use resources. Human agency is the capacity of human beings to make and exercise choices within their own cultural milieu. In this dissertation, I have developed a political ecology approach in order to show the interrelations and practices of people as they engage themselves in productive activities. In this sense, power is continually implicated and emergent in the inner workings of material social processes, material practices, and the outcomes of material struggles (Roseberry, 1989).

In this regard, the structural sources of insecurity that arise from social and climatic factors do not simply 'impact' the local pastoralists. Rather, these sources of vulnerability are clearly mediated by social problems of access and the distribution of power and agency in the dryland areas of Upper Awash Valley. As I have demonstrated clearly, people who are confronted with multiple sources of insecurity usually try to widen their range of options, i.e. to increase their flexibility so that they can achieve certain objectives. One way of doing this is through reactivation of traditional mechanisms of risk management and livelihood practices, such as mobility, and appealing to more than one single norm regarding access to and use of resources. This was clearly shown by the way the Karrayu pastoralists access resources by simultaneously claiming resources that help them engage both in cultivation and livestock rearing. Another way is through diversification (Barrett *et al.*, 2001). Karrayu pastoralists have diversified their activities, ranging from exclusive pastoralism that depends on camel herding based on inter-regional mobility, to intensified engagement in farming activities through arrangements with the Ittu farmers, and wage employment. The increased differentiation that exists within the Karrayu pastoral community today suggests that some pastoralists have been more successful than others in dealing with multiple sources of insecurity in the long term.

On the one hand, practices of appealing to more than one single norm regarding access to and use of scarce resources are closely related to wealth, so that the rich and moderately rich pastoralists can more easily resort to them. On the other hand, those without livestock often diversify their incomes spontaneously and in the short term, while wealthier pastoralists can afford to diversify their livelihood sources in a more strategic way. In other words, while some Karrayu pastoralists successfully embark upon a positive livelihood trajectory, others continually fail to do so. Another important conclusion that can be drawn from this dissertation is that the Karrayu pastoralists' interaction with the migrant Ittu farmers has a clear impact on resource decline through the conversion of pasture land to farm land. However, the impact of the influx of the Ittu to Karrayu land can be seen from two angles. On the one hand, with the migration and settlement of the Ittu in Karrayu territory came the increase of farming as an adaptation strategy by the Karrayu. The Ittu, with their prior knowledge of farming practices in their place of origin, brought the necessary farming skills that the Karrayu lack. Another aspect of this in-migration is the introduction of a new meaning to land, i.e. 'my land', in contrast to the communal ideas of the Karrayu. Through inter-marriage with the Karrayu, the Ittu farmers have comfortably located themselves in a position where their actions in many cases do not lead to conflict, while they are introducing the enclosure and privatization of land and thus putting in place new forms of resource management. As a result the communal resource base has declined, as well as the social orders and customary institutions that govern human-environment interactions. Interestingly, some informants explained local development in terms of the introduction of petty trading in the villages (providing access to various services), and agricultural intensification due to the increased population of the Ittu. On the other hand, some community members call for limitation of rapid population expansion.

The important issue here is how to make the link between rapid population growth and local adaptation strategies pursued by the Karrayu. In the light of the political ecology approach, the population growth of the farming Ittu is greater than the capacity of the land to support farming. This will seriously hamper the natural resource base and hence the livelihoods and food security of the Karrayu, who depend on the scarce resources to sustain their livelihoods. This would lead to the Neo-Malthusian trap. The competing theory proposed by Boserup (1965) holds that rapid population growth is one of the stimulant factors for the intensification of livelihoods. Although there are beginnings of intensification and diversification at both the study sites, it is too early to decide whether the Boserupian forecasts are right or not. However, the observations from the field suggest that the in-migration of non-locals into Karrayu territory has created pressure on the resources through the introduction of farming activities as a livelihood strategy. The large influx of Ittu farmers has also meant the introduction of Islam to the Karrayu community. Though this issue must be the topic of another project, its impact on the Karrayu pastoralists' mode of organization was noticed during this project. On the other hand, the migration of the Ittu to Karrayu territory has helped the smooth transfer of farming skills to the Karrayu herders who have already started to engage in farming but who have often failed due to lack of appropriate knowledge. All these processes of in-migration that took place at different periods, starting as early as the 1950s, show us how the interaction between active individual agents has produced and reproduced the present. The processes of in-migration of the Ittu to Karrayu land also reveal how sequences of change which are of medium-term duration can have far-reaching consequences for the region in question (Giddens, 1979:228).

8.3 LOCALIZING THE INTERNATIONAL PERSPECTIVE ON ADAPTATION TO CLIMATE CHANGE

By closely examining the climate change adaptation literature, particularly that regarding vulnerability-led approaches, and the wider vulnerability literature, particularly that stemming from disaster research, I have brought to light that despite a shift towards the concept of vulnerability in rhetoric, adaptation remains largely impact-focused. Such a focus on impacts is against the established theorization of the vulnerability-led approach in the disaster studies field. This narrow focus on impacts, I argue, reduces the role played by broader political, social and structural forces in adaptation. Even though these forces have little to do with climatic stimuli or climate change, they are the underlying forces behind people's vulnerability. In impact-focused approaches, adaptation is considered as additional to development and disaster risk reduction. In particular, I contend that this is to the impairment of effective pastoral adaptation, because it limits the ability to address vulnerability, which, as is revealed by objective one, is often caused by development-related failures.

This approach to vulnerability and adaptation reflects, and is sustained by, the wider mainstream climate change discourse which, I contend, is socially constructed and therefore partial. For this part of my review I adopted a critical realist perspective (see Chapter Two). I argue that the mainstream adaptation discourse is a product of a wider social construction of the climate change problem as biophysical and environmental. This construction of the climate change problem is based within a Western frame of reference, where nature and culture/society are separate entities and science and politics are disconnected. Climate change could equally be portrayed as a problem of development and inequality. However, in the mainstream literature and policy it is framed as a problem requiring mainly biophysical, environmental, apolitical solutions. From a constructivist perspective, I conclude that what constitutes vulnerability to climate change, and therefore, what actions are needed to adapt to it, are shaped by an inherently Western frame of thought in the mainstream adaptation discourse. This marginalizes the voices of 'others' in decision-making for adaptation.

Based on the case study of the Karrayu community in Upper Awash Valley, I conclude that for many pastoralists, vulnerability to climate stress is a primarily social, rather than a biophysical or 'environmental', phenomenon. While the mainstream adaptation discourse perpetuates an event-centred conceptual understanding of vulnerability to climate stress, the Karrayu pastoralists construct their own vulnerability as arising from a context of everyday lives and livelihoods. Climate stresses are not viewed as ontologically separate from society – they are not abnormal, external, or 'natural', but are a normal part of life and livelihood systems. Through local eyes, event-centred understandings of vulnerability are valid, but superficial. Rather than being the primary driver of vulnerability, climate stresses – like drought – merely unveil the social, cultural, economic and political factors that limit the ability to effectively respond to environmental uncertainty.

In the pastoral villages of Fentalle *Woreda*, vulnerability to climate is inseparable from development-related problems, which, to local communities, are a priority concern. At the core of resilience and adaptive capacity is the local, traditional way of organizing resources and people (*akka aadatti*). Localized pastoral mechanisms sustain, and enable the evolution of, local vulnerability reduction tools, allowing communities to adapt to environmental uncertainty. Many of these mechanisms are incidental features of society and livelihoods. However, socio-cultural change is reducing local adaptive capacity, separating vulnerability reduction from everyday life and livelihoods and increasing dependence on external resource flows. Processes of socio-cultural change are, through local eyes, at the core of increasing vulnerability to climate. The root causes of vulnerability, therefore, are distinctly development-related. Local people view problematic aspects of socio-cultural change as a product of state-sponsored development processes over time. 'Development' has eroded local self-sufficiency in vulnerability reduction by taking over the spaces of local adaptation, and has led to entitlement failure. At the same time, it has not provided higher scale safety nets to compensate this. The root causes of vulner- ability to climate, therefore, are largely outside the direct control of local communities.

I conclude that according to the local pastoralists in dryland areas of Upper Awash Valley, vulnerability to climate change is a political and ecological problem embedded in a wider political economy of development. Contrary to dominant constructions in the mainstream climate change adaptation discourse, Karrayu pastoralists' constructions of vulnerability resonate with the vulnerability paradigm in disaster research.

8.4 BACK TO THE RESEARCH QUESTIONS

In the preceding section I have outlined my major conclusions. Before finalizing, it is worth revisiting the research questions that guided the thesis in order to provide more concrete answers. The main question of the research was how have state interventions and the associated environmental transformations been experienced and acted upon by the local pastoral communities in the arid and semi-arid Metehara plains of Upper Awash Valley, Ethiopia? More specifically, the thesis answers the following questions:

A. What are the historical trajectories of insecurity that have influenced Karrayu pastoralists' risk management and livelihood practices?
B. How do Karrayu pastoralists continually practise livestock-based livelihood and risk management activities in the face of socio-environmental transformation?
C. How do Karrayu pastoralists take up and develop agro-pastoral livelihoods and risk management practices in the face of socio-environmental transformation?
D. How does research on context-specific risk management and livelihood security practices enable us to better reframe the mainstream perspective on adaptation?

The following six points represent, in a nutshell, the answers to these research questions:

Multiple sources of insecurity have influenced the Karrayu pastoralists. Unlike conventional and impact-oriented climate change adaptation research, which gives primacy to biophysical determinism, I have shown in this dissertation that when it comes to pastoralists the socio-political and demographic sources of insecurity play a central role in shaping the decision-making and adaptation strategies of the locals.

Pastoral actors make different responses to the various sources of insecurity. The local level is heterogeneous. There are various identities, encounters, strategies, and outcomes. The flexible risk management and livelihood practices depend on experience, current situation, and future ambitions.

Top-down adaptation may increase local vulnerability. Adaptation responses are often reactive and based on faith rather than a critical appraisal of the practices, needs and reality of the local environment. When this is the case, responses may reduce resilience and increase local vulnerability.

There is no 'one size fits all'. The present thesis has shown that the local people's responses are flexible and context-specific. Despite similar experiences that make them all equally vulnerable, the Karrayu pastoralists are following different livelihood trajectories. This is mainly because of the difference in pastoral agency that the locals possess and utilize. So, this needs to be acknowledged before embarking on any intervention to improve the local situation.

Karrayu pastoralists are creatively adapting to multiple vulnerabilities. Though the pressures from various sources of vulnerability are mounting, the Karrayu pastoralists, within the limits of their knowledge and other resources, are flexibly adapting by changing their behaviour and practices.

Adaptation strategies within the Karrayu community are diverse and vary across place. Despite the broadly similar local environment that Karrayu pastoralists dwell in, they try to spread their adaptation strategies across space and inscribe activities in different localities.

8.5 IMPLICATIONS FOR POLICY AND DIRECTIONS FOR FUTURE RESEARCH

The analysis of the pastoralists' risk management and livelihood practices from the perspective of a social vulnerability paradigm give us a better understanding of various local responses at the micro level. As I argued at the outset of this study, the way the local Karrayu pastoralists deal with multiple sources of insecurity can best be understood from a political ecology perspective, looking at how different pastoralists in the dryland parts of Upper Awash Valley recombine the old and the new, and improvise on practised routines in order to respond to particular challenges and opportunities arising from their broader environment. Therefore, a closer look at local actors, the norms and orders they refer to, the decisions they take and the practices they deploy improves our understanding of local adaptation, and thus of the long-term effects of human-environment interactions. However, the application of the concept of vulnerability in climate change adaptation research is in opposition to the field of disaster risk reduction research. Overemphasis on climate features, as in the case of the IPCC, diverts our attention from the development-related root causes that render the local Karrayu pastoralists vulnerable. By and large, the results presented in this study suggest the crucial role played by the social sources of vulnerability experienced by the Karrayu since the 1960s reflect the relevance of the vulnerability paradigm in disaster research. . In this context, many Karrayu pastoralists utilize their agency to deal with climate-related extreme events such as drought within the context of limited spaces of adaptation arising from the social sources of vulnerability. Therefore, localized adaptation to climate change in the Upper Awash Valley is a politico-ecological issue. At the core of the international climate change adaptation documents that guide national polices on adaptation, vulnerability is approached in terms of impacts. For instance, in the summary for policy makers, the IPCC document says that adaptation is essential to tackle impacts 'resulting from the warming' (IPCC, 2007: 19). The same document says that extensive adaptation is required to reduce vulnerability (IPCC, 2007). The premise of this document, which addresses policy makers, rests on the assumption that 'adaptation to climate change impacts can reduce vulnerability' (see Chapter Two for the connection between the concept of adaptation and vulnerability in climate change research). However, in my analysis, I have identified social sources of vulnerability that hamper local adaptation and thereby decrease their response to climate stress. Improving the pastoralists' situation in the dryland parts of Upper Awash Valley therefore means reducing the vulnerability arising from human interference, improving people's negotiating power, and thus helping them deal with climate stress better. This is based on a premise of disaster research where vulnerability reduction is seen as a facilitator of adaptation to climate change. One remaining question is how these negative social sources of vulnerability might eventually be dealt with.

8.5.1 Addressing social sources of vulnerability

Recently, a large number of development interventions in pastoral areas have addressed the problem of livelihood adaptation through the creation of new and 'stronger' institutions and organizations. Adddressing the social sources of vulnerability which usually emanate from lack of communal land security is vital. Communally based modes of relation to land on which the risk management and livelihood practise of the Karrayu highly depend need recognition. So understanding and addressing the social contexts in which risk management and livelihood practices operate should be part of the government's development interventions. Without such recognition interventions may not achieve the objective of creating resileint communities though their intentions are positive. One example is the government sedentarization programme for pastoralists, encouraging them by providing irrigated water. With such an institutionalized intervention, the development planners believe they can better deal with vulnerability. Such interventions build on experiences gained in a completely different context, i.e. a farming community. However, as Nuijten (2005: 5) comments, the limitation of these approaches often lies in their "faith that new forms of organizing and fresh rules can make a dramatic difference to the lives of the people and the management of resources. Official rules may influence existing organizing practices and power relations in many different and often unpredictable ways."

The situation among the Karrayu pastoral community underscores this. First, formal rules do not necessarily result in formal behaviour. Second, new regulations often coexist with older arrangements, thus adding to a hybrid institutional context that generally benefits already powerful actors. This does not mean that establishing new and better rules is meaningless. It does, however, mean recognizing that even policies with good intentions cannot avoid uncertainty and will always produce other sources of exposure. In line with this, Scoones (1995: 6) argues that "...an alternative for planning in an uncertain world is to accept the uncertainty and indeterminacy..."

As has been shown, vulnerability is usually more problematic for the poor, since they often lack the necessary negotiating power to make use of different norms and rules. At the same time, those with sufficient power are able to respond more flexibly to challenges and opportunities. Therefore, the fundamental question must be how the negotiating power of the poor can be reinforced.

8.5.2 Increasing the negotiation power of the locals

The general problem of top-down interventions in pastoral areas, such as large-scale irrigation schemes, lies in the belief that the provision of critical resources and privatization of farm land will create a level ground within the pastoral community. The presence of such resources may enhance a community's negotiating power in respect of state actors; however, it does not necessarily enhance the negotiating power of individuals within the pastoral community. The example of new forms of resource access that has been discussed in Chapter Six shows that powerful actors have a wide array of options in accessing scarce resources and can easily deal with climate-related stresses. Thus, any attempt at improving the Karrayu pastoralists' ability to deal with vulnerability in Upper Awash Valley must actively seek to strengthen the negotiating power of the community's poor and highly exposed members, by understanding the internal working of the pastoral community and devising rules that handle the differentially situated actors in a seemingly homogeneous community.

The substantiation presented in this study suggests that local processes of adaptation, including changes in social relations and individual organizing practices, are so multifaceted and flexible that it seems at least questionable whether a uniform approach (for example, through sedentarization) can do justice to every single Karrayu pastoralist. What this study therefore suggests is that, in the current situation, the introduction of new rules and regulations in rural parts of Fentalle *Woreda* needs cautious deliberation and must be rooted in a thorough understanding of pastoralists' ability to adapt and the locally specific processes that cause and reproduce disparities between potential stakeholders. Otherwise, apparently resilient rules and robust institutions run the risk of widening the existing gap between rich and poor.

8.5.3 Directions for future research

As has been noted by scholars such as Little (2013), pastoralism in east Africa is in a continual process of change and re-organization due to its increased integration into the national, regional and global political and economic system. This also applies to the Karrayu pastoral communities of upper Awash valley whose interaction with the Ethiopian state and wider socio-environment have triggered changes in their risk management and livelihood practices. First of all, it is to be expected that the implementation of the new irrigation scheme in the exclusively pastoral villages will initiate complex negotiations at various levels, depending on the meaning that pastoralists attach to the new resource introduced to their villages and also depending on the options that they have at hand. Even though access to water is one of the critical resources in the semi-arid environment of the Karrayu pastoralists and determines their ability to deal with climate-related sources of vulnerability, the distribution of previously communal land on an individual basis is creating discontent among the rich Karrayu pastoralists. It is clear that the provision of irrigation water can reduce the pressing biophysical impact of climate stresses. However, the long-term social implications of settlement and relocation of pastoralists for their vulnerability and resilience need to be further explored. Observing that process over the next few years and seeing how different groups of actors reposition themselves would certainly give new insights into the functioning of pastoral communities. In addition, the politics surrounding resources such as land and water, and the implications for inclusion and exclusion of individual pastoralists, need to be further researched. Second, the linkages between pastoral producers and agrarian commodity markets deserve more scientific attention. Thus, it would be interesting to examine how new forms of marketing could influence the use and management of the new irrigation scheme and the sedentarization of the Karrayu pastoralists.

Finally, with the arrival of irrigation in the Karrayu pastoral villages, a number of actors besides the Ittu farmers will be attracted to the Karrayu land. Such a dynamic process has the capacity to become the source of innovative local adaptation ideas, as well as conflicts based on how well the projective aspect of pastoral agency is utilized. In the absence of institutions that govern the changing situation, there is a high possibility of confrontation among various resource users in the future, with as yet unknown consequences for the livelihoods of the Karrayu pastoral population. Hence, understanding the implications of increased tension over privatized ownership for the social cohesion of the pastoralists and vulnerability-reduction mechanisms is important. In this regard, there is a need to understand the processes of institutional formation and how the old established customary institutions can be reinvigorated to fit into the new arrangements.

9. LIST OF REFERENCES

Adger, N. W., Lorenzoni, I., and O'Brien, K. (2009). Adaptation now. In N. W., Adger, I. Lorenzoni, and K. O'Brien (eds.), *Adapting to Climate Change: Thresholds, Values Governance* (pp. 1–23). New York: Cambridge University Press.

Adger, W. (2006). Vulnerability. *Global Environmental Change, 16*(3), 268–281.

Adger, W.N. and Brooks, N. (2003). Does global environmental change cause vulnerability to disaster? In M. Pellling (ed.), *Natural disasters and development in a globalising world* (pp. 19–42). London: Routledge.

Adger, W.N. and Kelly, P.M. (1999). Social vulnerability to climate change and the architecture of entitlements. *Mitigation and Adaptation Strategies for Global Change, 4*(3–4), 253–266.

Adger, W.N., Brooks, N., Bentham, G., Agnew, M. and Eriksen, S. (2004). *New indicators of vulnerability and adaptive capacity*. Tyndall Centre technical report no.7. Norwich: University of East Anglia.

Adger, W.N., Huq, S., Brown, K., Conway, D. and Hulme, M. (2003). Adaptation to climate change in the developing world. *Progress in development studies, 17*(1), 179–195.

Agrawal, A. (1999). *Greener Pastures: Politics, Markets, and Community among a Migrant Pastoral People*. Durham: Duke University Press.

Agrawal, A. (2005). *Environmentality: Technologies of Government and the Making of Subjects*. Durham: Duke University Press.

Allen, K. (2003). Vulnerability reduction and the community-based approach. In M. Pelling, (ed.), *Natural disasters and development in a globalizing world* (pp. 170–184). London, New York: Routledge.

Arce, A and Long, N. (2000). *Anthropology, development, and modernities: Exploring discourses, counter-tendencies, and violence*. London ; New York: Routledge.

Ayalew, G. (2001). *Pastoralism under pressure: Land Alienation and pastoral transformations among the Karrayu of Eastern Ethiopia, 1941 to the present*. Maastricht, Netherlands: Shaker Publishing.

Bailey, C. (2007). *A guide to qualitative field research*. London, Thousand Oaks, New Delhi: Sage Publications Ltd.

Bailey, F. G. (2001). *Stratagems and spoils: a social anthropology of politics*. Boulder, Colorado: Westview Press.

Bailey, S. and Bryant, R. (1997). *Third-World Political Eoclogy*. London: Routledge.

Bankoff, G. (2001). Rendering the world unsafe: 'vulnerability' as Western discourse. *Disasters, 25*(1), 19–35.

Bankoff, G. (2004). The historical geography of disaster: 'vulnerability' and 'local knowledge' in Western discourse. In G. F. Bankoff, G. Frerks, and D. Hilhorst (eds.), *Mapping vulnerability: disasters, development and people* (pp. 25–36). London, Sterling : Earthscan.

Barnett, J. (2010). Adapting to climate change: three key challenges for research and policy-an editorial essay. *Wiley Interdisciplinary Reviews: Climate Change, 1*(3), 314–317.

Barrett, C.B., Reardon, T. & Webb, P. (2001). Non-farm income diversification and household livelihood strategies in rural Africa: Concepts, dynamics, and policy implications. *Food Policy 26*, 315–331.

Berg, B. (2004). *Qualitative Research Methods for Social Science*. Boston, MA: Pearson publications.

Bernard, H. R. (1988). *Research methods in cultural anthropology*. Newbury Park, California: Sage Publications.

Bernard, H. R. (1995). *Research methods in anthropology: Qualitative and quantitative methods (2nd ed.).* Walnut Creek, CA: AltaMira Press.
Berry, S. (1989). Social Institutions and Access to Resources. *Africa, 59*(1), 41–55.
Berry, S. (1993). *No Condition is Permanent: The Social Dynamics of Agrarian Change in Sub-Saharan Africa.* Madison: University of Wisconsin Press.
Best, U. (2009). Critical geography. In N. Thrift and R. Kitchin (eds.), *International encyclopedia of human geography* (pp. 345–357). Oxford: Elsevier Science.
Biersack, A. and Greenberg, J. B. (2006). *Reimagining political ecology.* Durham: Duke University Press.
Blaikie, P. (2008). Epilogue: Towards a future for political ecology that works. *Geoforum, 39*(2), 765–772.
Blaikie, P. and Brookfield, H. (1987). *Land degradation and society.* London, New York: Methuen.
Bohle, H.G., Downing, T.E. and Watts, M.J. (1994). Climate change and social vulnerability: toward a sociology and geography of food insecurityl. *Globa Environmental Change 4*, 37–48.
Boko, M., I. Niang, A. Nyong, C. Vogel, A. Githeko, M. Medany, B. Osman-Elasha, R. Tabo and P. Yanda. (2007). Africa. In M.L. Parry, O.F. Canziani, J.P. Palutikof, P.J. van der Linden and C.E. Hanson (eds.), *Climate Change 2007: Impacts, Adaptation and Vulnerability. Contribution of Working Group II to the Fourth Assessment Report of the Intergovernmental Panel on Climate Change* (pp. 433–467). Cambridge, UK: Cambridge University Press.
Bollig, M. (2006). *Risk Management in a Hazardous Environment: A Comparative Study of Two Pastoral Societies.* New York: Springer and Business Media.
Bollig, M. and Göbel, B. (1997). Risk, Uncertainty and Pastoralism: An Introduction. *Nomadic Peoples, 1*(1), 5–21.
Bondestam, L. (1974). Peoples and Capitalism in the Northeast Lowlands of Ethiopia. *Journal of Modern African Studies, 12*, 428–439.
Boserup, E. (1965). *The Conditions of Agricultural Growth.* New York: Aldine.
Brooks, N. (2003). *Vulnerability, risk and adaptation:a conceptual framework. Tyndall Centre working paper no. 38.* Norwich: University of East Anglia.
Brown, K and Westaway, E. (2011). Agency, capacity, and resilience to environmental change: lessons from human development, well-being, and disasters. *Annual review of environment and resources, 36*(1), 321–342.
Brown, K. and Westaway, E. (2011). Agency, capacity, and resilience to environmental change: lessons from human development, well-being, and disasters. *Annual review of environment and resources, 36*(1), 321–342.
Bryant, R. (1998). Power, Knowledge and Political Ecology in the Third World: a review. *Progress in Physical Geography, 22*(1), 79–94.
Burton, I. (2009). Deconstructing adaptation ... and reconstructing. In L. Schipper and I. Burton (eds.), *The Earthscan reader on adaptation to climate change* (pp. 11–14). London, Sterling: Earthscan.
Burton, I., Huq, S., Lim, B., Pilifosova, O. and Shipper, L. (2002). From impacts assessment to adaptation priorities: the shaping of adaptation policy. *Climate Policy, 2*, 145–159.
Burton, I., Kates, R. and White, G. (1978). *The environment as hazard.* Oxford: Oxford University Press.
Carter, T.R., Jones, R.N., Lu, X., Bhadwal, S., Conde, C., Mearns, L.O., O'Neill, B.C., Rounsevell, M.D.A. and Zurek, M.B. (2007). New assessment methods and the characterisation of future conditions. In M. C. Parry, *Climate change 2007: impacts, adaptation and vulnerability. Contribution of Working Group II to the Fourth Assessment Report of the Intergovernmental Panel on Climate Change* (pp. 133–171). Cambridge: Cambridge University Press.

Cass, L.R. and Pettenger, M.E. (2007). Conclusion: the constructions of climate change. In M.E. Pettenger (ed.), *The social construction of climate change: power, knowledge, norms, discourses* (pp. 235–246). Aldershot, Burlington: Ashgate Publishing Ltd.

Catley Andy, Lind Jeremy, Scoones Ian (Eds.). (2013). *Pastoralism and Development in Africa: Dynamic Change at the Margins.* London: Routledge.

Chambers, R. (1983). *Rural Development: Putting the last first.* Harlow: Prentice Hall.

Chambers, R. (1989). Vulnerability, coping and policy. *Institute of Development Studies Bulletin, 20*(2), 1–7.

Clapham, C. (2009). Post War Ethiopia: Trajectories of Crisis. *Review of African Political Economy, 36*(120), 181–192.

Clifford, N.J and Valentine, G. (2003). Getting started in geographical research: how this book can help. In N. Clifford and G. Valentine (eds.), *Key methods in geography* (pp. 1–16). Thousand Oaks, London, New Delhi: Sage Publications Ltd.

Cook, I. (1997). Participant observation. In R. Flowerdew and D. Martin (eds.), *Methods in human geography* (pp. 167–188). London: Harlow.

Cornwall, A. and Brock, K. (2006). What do buzzwords do for development policy? a critical look at 'participation', 'empowerment' and 'poverty reduction'. *Third World Quarterly, 26(7), 1043–1060.*

Creswell, J. W. (1994). *Research Design: Qualitative and Quantitative Approaches. First Edition.* Thousand Oaks, CA: Sage Publications.

Creswell, J. W. (2003). *Research Design: Qualitative, Quantitative, and Mixed Methods Approaches. Second Edition.* Thousand Oaks, CA: Sage Publication.

Creswell, J. W. (2007). *Qualitative Inquiry and Research Design: Choosing Among Five Approaches (2 ed.).* Thousand Oaks, London: New Delhi Sage Publications Ltd.

Creswell, J. W. (2007). *Qualitative Inquiry and Research Design: Choosing Among Five Approaches. Second Edition.* Thousand Oaks, CA: Sage Publication.

Cutter, S. (1996). Vulnerability to environmental hazards. *Progress in Human Geography 20,* 529–539.

Davidson, O., Halsnæsb, K., Huq, S., Kok, M., Metz, B., Sokona, Y. and Verhagenf, J. (2003). The development and climate nexus: the case of sub-Saharan. *Africa Climate Policy, 3*(1), 97–113.

Davies, J and Bennett, R. (2007). Livelihood Adaptation to risk: Constraints and opportunities in Ethiopia's Afar region. *The Journal of development Studies, 43*(3), 490–511.

Davis, I. (2004). Progress in analysis of social vulnerability and capacity. In G. F. Bankoff, *Mapping vulnerability: disasters, development and people* (S. 128–144). London, Sterling: Earthscan.

Demeritt, D. (2001). The construction of global warming and the politics of science. *Annals of the Association of American Geographers, 91*(2), 307–337.

Demeritt, D. (2006). Science studies, climate change and the prospects for constructivist critique. *Economy and Society, 35*(3), 453–479.

Dessai, S., Adger, W.N., Hulme, M., Turnpenny, J.R., Kohler, J. and Warren, R. (2004). Defining and experiencing dangerous climate change. *Climatic Change, 64*(1), 11–25.

Dowling, R. (2005). Power, subjectivity, and ethics in qualitative research. In I. Hay (ed.), *Qualitative research methods in human geography (2nd ed.)* (pp. 19–29). Melbourne: Oxford University Press.

Dunn, K. (2005). Interviewing. In I. Hay (ed.), *Qualitative research methods in human geography (2nd ed.)* (pp. 79–105). Melbourne: Oxford University Press.

Dyer, C. (2013). Does mobility have to mean being hard to reach? Mobile pastoralists and educaion's 'terms of inclusion'. *Compare: A Journal of Comparative and International Education, 43*(5), 601–621.

Elias, E. (2008). *Pastoralists in Southern Ethiopia: Dispossession, Access to Resources and Dialogue with Policy Makers.* Drylands Coordination Group. Report No. 53.

Emerson, R.M, Fretz, R.I, and Shaw, L.L. (1995). *Writing ethnographic fieldnotes*. Chicago: University of Chicago Press.
Emirbayer, M. and Mische, A. (1998). What is Agency? *The American Journal of Sociology, 103*(4), 962–1023.
Ensminger, J. (1996). *Making a Market: The Institutional Transformation of an African Society*. Cambridge: Cambrdige University Press.
Ensminger, J. and Andrew R. (1990). The political economy of changing property rights: dismantling a pastoral commons. *American Ethnologist*, 683–699.
Ensor, J. and Berger, R. (2009). Introduction: understanding community-based adaptation. In J. Ensor and R. Berger (eds.), *Understanding climate change adaptation: lessons from community-based approaches* (pp. 3–20). Rugby: Practical Action Publishing Ltd.
Eriksen, S., & Lind, J. (2009). Adaptation as a Political Process: Adjusting to Drought and Conflict in Kenya's Drylands. *Environmental Management, 43*(5), 817–835.
Eriksen, S., Aldunce, P., Bahinipati, S. C., Martins, D. R., Molefe, I. J., Nhemachena, C., O'brien, K., Olorunfemi, F., Park, J., Sygna, L and Ulsrud, K. (2011). When not every response to climate change is a good one: Identifying principles for sustainable adaptation. *Climate and Development, 3*(1), 7–20.
Eriksen, S., Brown, K. and Kelly, P.M. (2005). The dynamics of vulnerability: locating coping strategies in Kenya and Tanzania. *The Geograhical Journal, 171*(4), 287–385.
Eriksen, S.H. and Kelly, P.M. (2007). Developing credible vulnerability indicators for climate adaptation policy assessment. *Mitigation and Adaptation Strategies for Global Change 12*, 495–524.
Escobar, A. (1999). After Nature: Steps to an Anti-essentialist Political Ecology. *Current Anthropology, 40*(1), 1–30.
Flintan, F. (2011). The Political Economy of Land Reform in Pastoral Areas: Lessons from Africa, Implications for Ethiopia. *International Conference on the Future of Pastoralism, 21–23 March 2011*.
Forsyth, T. (2003). *Critical political ecology: the politics of environmental science*. London, New York: Routledge.
Fussel, H. M. (2007). Vulnerability: a generally applicable conceptual framework for climate change research. *Global Environmental Change, 17*(2), 155–167.
Fussel, H.M. and Klein, R. (2006). Climate change vulnerability assessments: an evolution of conceptual thinking. *Climatic Change 75*, 301–329.
Gaillard, J. (2010). Vulnerability, capacity and resilience: perspectives for climate and development policy. *Journal of International Development 22*, 218–232.
Gerring, J. (2007). *Case study research: principles and practices*. Cambridge, Massachusetts: Cambridge University Press.
Getachew, K. (2001). *Among the pastoral Afar in Ethiopia: Tradition, continuity and Socio-Economic change*. Utrecht: International Books.
Gezon, L. L. (2006). *Global visions, local landscapes: a political ecology of conservation, conflict, and control in Northern Madagascar*. Lanham, MD: AltaMira Press.
Giddens, A. (1976). *New rules of sociological method : a positive critique of interpretative sociologies*. New York: Basic Books.
Giddens, A. (1979). *Central problems in social theory: Action, structure, and contradiction in social analysis*. Berkeley: University of California Press.
Giddens, A. (1984). *The constitution of society: Outline of the theory of structuration*. Cambridge: Polity Press.
Greenberg, J.B and Park, T.K. (1994). Political ecology. *Journal of Political Ecology 1*, 1–12.
Hagmann, T. and Alemmaya, M. (2008). Pastoral Conflicts and State-building in the Ethiopian Lowlands. *Africa Spectrum, 43*(1), 19–37.
Hammersley, M. and Atkinson, P. (1983). *Ethnography: Principles in Practice*. London: Routledge.

Heijmans, A. (2004). From vulnerability to empowerment . In G. F. Bankoff, G. Frerks, and D. Hilhorst (eds.), *Mapping vulnerability: disasters, development and people* (pp. 115–127). London, Sterling: Earthscan.

Helland, J. (2006). Land Tenure in the Pastoral Areas of Ethiopia. *International Research Workshop on Property Rights, Collective Action and Poverty Reduction in Pastoral Areas of Afar and Somali National Regional State, Ethiopia, 30–31 October, 2006.*

Hewitt, K. (1983a). The idea of calamity in a technocratic age. In K. Hewitt (ed.), *Interpretations of calamity: from the viewpoint of human ecology* (pp. 3–32). Boston: Allen and Unwin.

Hewitt, K. (1983b). *Interpretations of calamity: from the viewpoint of human ecology.* Boston: Allen and Unwin.

Hickey, S. and Kothari, U. (2009). Participation. In N. Thrift and R. Kitchin (eds.), *International encyclopedia of human geography* (pp. 82–89). Oxford: Elsevier.

Huq, S. and Reid, H. (2004). Mainstreaming adaptation in development. *Institute for Development Studies Bulletin 35*, 15–21.

Huq, S., Rahman, A., Konate, M., Sokona, Y. and Reid, H. (2003). *Mainstreaming adaptation to climate change in Least Developed Countries (LDC's).* London: IIED.

IPCC. (2007). Summary for Policymakers. In M.L. Parry, O.F. Canziani,J.P. Palutikof, P.J. van der Linden and C.E. Hanson (eds.), Climate Change 2007: Impacts, Adaptation and Vulnerability. Contribution of Working Group II to the Fourth Assessment Report of the Intergovernmental Panel on Climate Change (pp. 7–22). Cambridge, UK: Cambridge University Press.

IPCC. (2014). Summary for policymakers. In C.B., Barros, D.J. Dokken, K.J. Mach, M.D. Mastrandrea T.E. Bilir, M. Chatterjee, K.L. Ebi, Y.O. Estrada, R.C. Genova, B. Girma, E.S. Kissel, A.N. Levy, S. MacCracken P.R. Mastrandrea, and L.L.White (eds.), *Climate Change 2014: Impacts,Adaptation, and Vulnerability Part A: Global and Sectoral Aspects. Contribution of Working Group II to the Fifth Assessment Report of the Intergovernmental Panel on Climate Change* (pp. 1–32). Cambridge, United Kingdom and New York, USA: Cambridge University Press.

Johnson, R. B. and Onwuegbuzie, A. J. (2004). Mixed methods research: A research paradigm whose time has come. *Educational Researcher, 33*(7), 14–26.

Kearns, R. (2005). Knowing seeing? Undertaking observational research. In I. Hay (ed.), *Qualitative research methods in human geography (2nd ed.)* (pp. 192–206). Melbourne: Oxford University Press.

Kelly, P.M. and Adger, W. N. (2000). Theory and practise in assessing vulnerability to climate change and facilitating adaptation. *Climate Change 47*, 325–352.

Kerkvliet, B. J. (2009). Everyday politics in peasant societies (and ours). *The Journal of Peasant Studies, 36*(1), 227–243.

Kindon, S. (2005). Participatory action research. In I. e. Hay, *Qualitative research methods in human geography (Second edition)* (S. 207–220). Melbourne: Oxford University Press.

Kitchin, R. and Tate, N.J. (2000). *Conducting Research into Human Geography: Theory, Methodology and Practise.* Essex: Pearson Education Limited.

Klein, R., Schipper, L. and Dessai, S. (2003). Integrating mitigation and adaptation into climate and development policy: three research questions. *Tyndall Centre working paper no. 40. Norwich. University of East Anglia.*

Lane, C. and Moorehead, R. (1995). New Directions in Rangeland Resource Tenure and Policy. In I. Scoones (ed.), *Living with Uncertainty: New Directions in Pastoral Development in Africa* (pp. 116–133). London: Intermediate Technology Publications.

Latour, B. (2004). *Politics of Nature: How to bring the sciences back into democracy.* Cambridge: Harvard University Press.

Leach, M, Mearns, R., and Scoones, I. (1999). Environmental Entitlements: Dynamics and Institutions in Community-based Natural Resource Management. *World Development, 27*(2), 225–247.

Little, P. (2013). Reflections on the future of pastorlism in the Horn of Africa. In A. Catley, J. Lind, and I. Scoones (eds.), *Pastoralism and development in Africa: Dynamic change at the margins* (pp. 243–249). London: Routledge and Earthscan.

Liverman, D. (2009). Conventions of climate change: constructions of danger and the dispossession of the atmosphere. *Journal of Historical Geography, 35*(2), 279–295.

Long, N. and Long, A. (1992). *Battlefield of Knowledge: The Interlocking of Theory and Practice in Social Research and Development.* London and New York: Routledge.

Markakis, J. (2011). *Ethiopia: The Last Two Frontiers.* London: James Currey.

Mayoux, L. (2006). Quantitative, qualitative or participatory? Which method, for what and when? In V. Desai, and R.B. Potter (eds.), *Doing development research* (pp. 115–129). Thousand Oaks, London, New Delhi: Sage Publications Ltd.

McCabe, T. (1994). *African Pastoralist Systems: An Integrated Approach.* Boulder, Colorado: Lynne Rienner Publishers.

McCabe, T. (2004). *Cattle Bring Us to Our Enemies: Turkana Ecology, Politics, and Raiding in a Disequilibrium System.* Ann Arbor: The University of Michigan Press.

McGray, H., Hammill, A. and Bradley, R. (2007). *Weathering the storm: options for framing adaptation and development.* Washington D.C.: World Resources Institute.

McLaughlin, P. and Dietz, T. (2008). Structure, agency and environment: Toward an integrated perspective on vulnerability. *Global Environmental Change, 18*(1), 99–111.

Morton, J. (2010). Why should governmnetality matter for the study of pastoral development? *Nomadic Peoples, 14*(1), 6–30.

Muderis, A. (1998). *Resource Deprivation and Socio-economic changes among Pastoral Households: The case of Karrayyu and Ittu Pastoralists in the Middle Awash valley of Ethiopia.* M.A. Thesis. Oslo: Agricultural University of Norway.

Müller-Mahn, D., Girum Getachew, and Rettberg S. (2010). Pathways and Dead Ends of Pastoral Development among the Afar and Karrayu in Ethiopia. *European Journal of Development Research, 22*, 660–677.

Nelson, D. and Finan, T. (2009). Praying for Drought: Persistent Vulnerability and the Politics of Patronage in Northeast Brazil. *American Anthropologist, 111*(3), 302–316.

Neumann, R. P. (1998). *Imposing Wilderness: Struggles over livelihood and nature preservation in Africa.* Berkeley and Los Angeles: University of California Press.

Nuijten, M. (2005). Power in practice: a force field approach to natural resource management. *The Journal of Transdisciplinary Environmental Studies, TES Special Issue, 4*(2).

O'Keefe, P., Westgate, K. and Wisner, B. (1976). Taking the naturalness out of natural disasters. *Nature, 260*(5552), 566–567.

O'Brien, G., O'Keefe, P., Rose, J. and Wisner, B. (2006). Climate Change and Disaster Management. *Disasters, 30*(1), 64–80.

O'Brien, K and Leichenko, R. (2000). Double exposure: assessing the impacts of climate change within the context of economic globalization. *Global Environmental Change, 10*, 221–232.

O'Brien, K, Eriksen, S, Schjolden, A and Nygaard, L. (2004). What's in a word? Conflicting interpretations of vulnerability in climate change research. *CICERO working paper 2004:04*.

O'Brien, K. and St. Clair, A.L. (2007). Shifting the discourse: climate change as an issue of human security. *European Science Foundation Exploratory Workshop Proceedings, 22–23 June.* Oslo: European Science Foundation.

Oliver-Smith, A. (2004). Theorizing vulnerability in a globalized world: a political ecological perspective. In G. F. Bankoff, G. Frerks and D. Hilhorst (eds.), *Mapping vulnerability: disasters, development and people* (pp. 10–24). London, Sterling: Earthscan.

Olmos, S. (July 2001). *Vulnerability and adaptation to climate change: concepts, issues, assessment methods.* Climate Change Knowledge Network.

Overton, J. and van Dierman, P. (2003). Using quantitative techniques. In R. Scheyvens and and D. Storey (eds.), *Development fieldwork: a practical guide* (pp. 37–56). London, Thousand Oaks, New Delhi: Sage Publications.

Parry, M. and Carter, T. (1998). *Climate impact and adaptation assessment: a guide to the IPCC approach.* London: Earthscan.

Patton, M. (2002). *Qualitative Research and Evaluation Methods (3rd ed.).* Thousand Oaks, London, New Delhi: Sage Publications Ltd.

Paulson, S. and Gezon, L. L, Eds. (2005). *Political ecology across spaces, scales, and social groups.* New Brunswick: Rutgers University Press.

Pielke, R. (1998). Rethinking the role of adaptation in climate policy. *Global Environmental Change, 8*(2), 159–170.

Pielke, R. (2005). Misdefining "climate change": Consequences for science and action. *Environmental Science and Policy 8*, 548–561.

Pittock, B. and Jones, R. (2009). Adaptation to what and why? In L. Shipper and I. Burton (eds.), *The Earthscan reader on adaptation to climate change* (pp. 35–62). London, Sterling: Earthscan.

Pretty, J. N. (1994). Alternative systems of inquiry for sustainable agriculture. *IDS Bulletin , 25*(2), 37–48.

Pretty, N. & Vodouhê, D. (1997). Using rapid or participatory rural appraisal. In E. B. Swanson, P. Bentz, J. and Sofranko (eds.), *Improving agricultural extension: a reference manual.* Rome: Food and Agriculture Organization of the United Nations.

Reid, P. and Vogel, C. (2006). Living and responding to multiple stressors in South Africa-Glimpses from KwaZulu-Natal. *Global Environmental Change, 16*(2), 195–206.

Ribot, J. (1998). Theorizing Access: Forest Profits along Senegal"s Charcoal Commodity Chain. *Development and Change, 29*(2), 307–341.

Ribot, J. and Peluso, N. (2003). A Theory of Access. *Rural Sociology, 68*(2), 153–181.

Robbins, P. (2012). *Political ecology: a critical introduction.* West Sussex: Blackwell Publishing Ltd.

Rocheleau, D. (1995). Maps, Numbers, Text, and Context: Mixing Methods in Feminist Political Ecology. *Professional Geographer , 47*(4), 458–466.

Rocheleau, D. Thomas-Slayter, B and Wangari, E., Eds. (1996). *Feminist political ecology: Global issues and local experience. International studies of women and place.* London: Routledge.

Roseberry, W. (1989). *Anthropologies and Histories: Essays in Culture, History, and Political Economy.* New Brunswick: Rutgers University Press.

Schipper, L. and Burton, I. (2009). Understanding adaptation: origins, concepts, practice and policy. Earthscan reader on adaptation to climate change. In L. Schipper and I. Burton (eds.), *The The Earthscan reader on adaptation to climate change* (pp. 1–10). London, Sterling. Earthscan.

Schipper, L. and Pelling, M. (2006). Disaster risk, climate change and international development: scope for and challenges to integration.,. *Disasters, 30*(1), 19–38.

Schipper, L. (2007). *Climate change adaptation and development: exploring the linkages.* Tyndall Centre working paper no. 107. Norwich: University of East Anglia.

Schipper, L. (2009). Meeting at the crossroads?: exploring the linkages between climate change adaptation and disaster risk reduction. *Climate and Development, 1*(1), 16–30.

Scoones, I. (1995). New directions in pastoral development in Africa. In I. Scoones (ed.), *Living with uncertainty: new directions in pastoral development in Africa* (pp. 1–36). London: Institute of Development Studies.

Scott, J. (1985). *Weapons of the Weak: Everyday Forms of Resistance.* New Haven: Yale University Press.

Scott, J. (1998). *Seeing like a State: how certain schemes to improve the human condition have failed.* New Haven: Yale University Press.

Sen, A. (1981). *Poverty and famines: an essay on entitlement and deprivation.* Oxford: Oxford University Press.

Silva, J.A., Eriksen, S. and Ombe, Z.A. (2010). Double exposure in Mozambique's Limpopo River Basin. *Geographical Journal 176*, 6–24.

Smit, B. and Pilifosova, O. (2001). Adaptation to climate change in the context of sustainable development and equity. In J. McCarthy, O.F. Canziani, N.A. Leary, D.J. Dokken, and K.S White (eds.), *Climate Change 2001: impacts, adaptation and vulnerability. Contribution of Working Group II to the Third Assessment Report of the Intergovernmental Panel on Climate Change.* Cambridge: Cambridge University Press.

Smit, B. and Pilifosova, O. (2003). From adaptation to adaptive capacity and vulnerability reduction. In J. Smith, R.T.J. Klein, and S. Huq (eds.), *Climate change, adaptive capacity and development* (pp. 9–28). London: Imperial College Press.

Smit, B. and Wandel, J. (2006). Adaptation, adaptive capacity and vulnerability. *Global Environmental Change, 16*(3), 282–292.

Smit, B., Burton, I., Klein, R. and Wandel, J. (2000). An anatomy of adaptation to climate change and variability. *Climatic Change, 45*(1), 223–251.

Sokona, Y. and Huq, S. (2002). *Climate change and sustainable development: views from the south. World Summit on Sustainable Development opinion paper.* London: IIED.

Spencer, P. (1997). *The Pastoral Continuum: The Marginalization of Tradition in East Africa.* Oxford: Oxford Univeristy Press.

Stake, R. (2005). Qualitative case studies. In K. Denzin, and S. Lincoln (eds.), *The Sage handbook of qualitative research* (3rd ed.). London: Sage.

Swidler, A. (1986). Culture in action: Symbols and strategies. *American sociological review*, 273–286.

Tache, B and Oba, G. (2009). Policy-driven Inter-ethnic Conflicts in Southern Ethiopia. *Review of African Political Economy , 38*(121), 409–426.

Tsing, A. L. (2005). *Friction: An Ethnography of Global Connection.* Princeton: Princeton University Press.

Turner, B.L. and Robbins, P. (2008). Land-change science and political ecology: similarities, differences, and implications for sustainability science. *Annual Review of Environmental Resources 33*, 295–316.

Turner, B.L., Kasperson, R.E., Matson, P.A., McCarthy, J.J., Correll, R.W., Christiensen, L., Eckley, N., Kasperson, J.X., Luers, A., Martello, M.L., Polsky, C., Pulsipher, A. and Shciller, A. (2003). A framework for vulnerability analysis in sustainability science. *Proceedings of the National Academy of Sciences 100*, 8074–8079.

UN. (1992). *United Nations Framework Convention on Climate Change.* Geneva: United Nations.

UN. (1992). *United Nations Framework Convention on Climate Change.* Geneva: United Nations.

UNFCCC. (2002). *Annotated guidelines for the preparation of national adaptation programmes of action. Least Developed Countries expert group.* UNFCCC.

UNFCCC. (2005). *Compendium on methods and tools to evaluate impacts of, and vulnerability and adaptation to, climate change.* Bonn Secretariat: UNFCCC.

Watts, M. (1983). On the poverty of theory: natural hazards research in context. In K. Hewitt (ed.), *Interpretations of calamity from the viewpoint of human ecology* (pp. 231–262). Boston: Allen and Unwin.

Watts, M.J. and Bohle, H.G. (1993). The space of vulnerability: the causal structure of hunger and famine. *Progress in Human Geography, 17*(1), 43–67.

Wilbanks, T. (2003). Integrating climate change and sustainable development in a place-based context. *Climate Policy, 3*(1), 147–154.

Willis, J. (2007). *Foundations of qualitative research .* Thousand Oaks, London, New Delhi: Sage Publications Ltd.

Wisner, B., Blaikie, P., Cannon, T. and Davis, I. (2004). *At risk: natural hazards, people's vulnerability and disasters (2nd ed.).* London: Routledge.

Yamin, F. (2004). Climate change and development: overview. *Institute of Development Studies Bulletin, 35*(3), 1–10.

Yohe, G.W., Lasco, R.D., Ahmad, Q.K., Arnell, N.W., Cohen, S.J., Hope, C., Janetos, A.C. and Perez, R.T. (2007). Perspectives on climate change and sustainability. In M. C. Parry, M.L., Canziani, O.F., Palutikof, J.P., van der Linden, P.J. and Hanson, C.E. (eds.), *Climate change 2007: impacts, adaptation and vulnerability. Contribution of Working Group II to the Fourth Assessment Report of the Intergovernmental Panel on Climate Change* (pp. 811–841). Cambridge: Cambridge University Press.

Zimmerer, K and Bassett, T. (2003). *Political Ecology: An Integrative Approach to Geography and Environment-Development Studies.* New York: Guilford Publications.

ERDKUNDLICHES WISSEN
Schriftenreihe für Forschung und Praxis

Begründet von Emil Meynen.
Herausgegeben von Martin Coy, Anton Escher und Thomas Krings.

Franz Steiner Verlag   ISSN 0425-1741

145. Heike Egner
Gesellschaft, Mensch, Umwelt – beobachtet
Ein Beitrag zur Theorie der Geographie
2008. 208 S. mit 8 Abb., 1 Tab., kt.
ISBN 978-3-515-09275-3

146. *in Vorbereitung*

147. Heike Egner, Andreas Pott
Geographische Risikoforschung
Zur Konstruktion verräumlichter Risiken und Sicherheiten
2010. XI, 242 S. mit 16 Abb., 3 Tab., kt.
ISBN 978-3-515-09427-6

148. Torsten Wißmann
Raum zur Identitätskonstruktion des Eigenen
2011. 204 S., kt.
ISBN 978-3-515-09789-5

149. Thomas M. Schmitt
Cultural Governance
Zur Kulturgeographie des UNESCO-Welterberegimes
2011. 452 S. mit 60 z.T. farb. Abb., 17 Tab., kt.
ISBN 978-3-515-09861-8

150. Julia Verne
Living Translocality
Space, Culture and Economy in Contemporary Swahili Trade
2012. XII, 262 S. mit 45 Abb., kt.
ISBN 978-3-515-10094-6

151. Kirsten von Elverfeldt
Systemtheorie in der Geomorphologie
Problemfelder, erkenntnistheoretische Konsequenzen und praktische Implikationen
2012. 168 S. mit 13 Abb., kt.
ISBN 978-3-515-10131-8

152. Carolin Schurr
Performing Politics, Making Space
A Visual Ethnography of Political Change in Ecuador
2013. 213 S. mit 36 Abb., 2 Ktn. und 10 Tab., kt.
ISBN 978-3-515-10466-1

153. Matthias Schmidt
Mensch und Umwelt in Kirgistan
Politische Ökologie im postkolonialen und postsozialistischen Kontext
2013. 400 S. mit 26 Abb., 12 Tab. und 16 Farbtafeln mit 8 Fotos und 12 Karten, kt.
ISBN 978-3-515-10478-4

154. Andrei Dörre
Naturressourcennutzung im Kontext struktureller Unsicherheiten
Eine Politische Ökologie der Weideländer Kirgisistans in Zeiten gesellschaftlicher Umbrüche
2014. 416 S. mit 29 Abb., 14 Tab. und 35 Farbabb. auf 24 Taf., kt.
ISBN 978-3-515-10761-7

155. Christian Steiner
Pragmatismus – Umwelt – Raum
Potenziale des Pragmatismus für eine transdisziplinäre Geographie der Mitwelt
2014. 290 S. mit 9 Abb., 7 Tab., kt.
ISBN 978-3-515-10878-2

156. Juliane Dame
Ernährungssicherung im Hochgebirge
Akteure und ihr Handeln im Kontext des sozioökonomischen Wandels in Ladakh, Indien
2015. 368 S. mit 49 s/w- und 28 Farbabb., 4 Farb- und 2 s/w-Karten sowie 2 farbigen Faltkarten, kt.
ISBN 978-3-515-11032-7

157. Jürg Endres
Rentierhalter. Jäger. Wilderer?
Praxis, Wandel und Verwundbarkeit bei den Dukha und den Tozhu im mongolisch-russischen Grenzraum
2015. 452 S. mit 12 s/w-Abb., 4 Tab., 53 s/w-Fotos sowie 12 Farb- und 10 s/w-Karten, kt.
ISBN 978-3-515-11140-9

158. Angelo Gilles
Sozialkapital, Translokalität und Wissen
Händlernetzwerke zwischen Afrika und China
2015. 265 S. mit 6 Abb. und 2 Tab., kt.
ISBN 978-3-515-11169-0